FOREWORD

OFFICE OF THE ASSISTANT SECRETARY OF DEFENSE
2500 DEFENSE PENTAGON
WASHINGTON, D.C. 20301-2500

SPECIAL OPERATIONS/
LOW-INTENSITY CONFLICT

The U.S. Government (USG) responds to approximately 70-80 natural disasters across the globe each year. In approximately 10-15 percent of these disaster responses, the Department of Defense (DoD) lends support to the overall USG effort. In these instances, DoD acts in support of the Department of State (DoS) and the United States Agency for International Development (USAID), and in close coordination with the affected country and the international humanitarian community, including other donor countries and participating international organizations and non-governmental organizations.

DoD disaster assistance can range from a single aircraft delivering relief supplies, to a full-scale deployment of a brigade-size or larger task force. Though the overall percentage of disasters requiring DoD support is relatively small, these disasters tend to be crises of the largest magnitude and/or the greatest complexity. This is largely due to the unique contribution our military can make to these types of responses. Capabilities such as strategic airlift and expeditionary medical care and engineering enable the USG to provide quick support to affected countries in the immediate aftermath of a devastating disaster. DoD assets are designed to operate in austere conditions, enabling them to contribute in post-disaster environments where critical infrastructure has been severely damaged or destroyed. Furthermore, the military's expertise in logistics can prove to be an invaluable tool for mobilizing large amounts of assistance in a short period of time.

Over the past decade, DoD disaster relief support has included the response to Southeast Asia following the tsunami in 2004 and assistance to Pakistan after a massive earthquake in 2005 and flooding in 2010. In January 2010, DoD partnered with other U.S. departments and agencies to support the largest USG disaster response in history after Haiti's catastrophic earthquake. More than 20,000 U.S. military personnel were deployed in support of Operation UNIFIED RESPONSE, executing a wide range of duties including reopening the airport and supporting airport operations, providing security for relief operations, conducting rotary and fixed wing sorties to transport relief commodities and personnel, providing engineering support to displaced persons camps, repairing critical infrastructure at the seaport, and distributing relief supplies.

In the aftermath of the Haiti earthquake response, DoD's after-action reviews and lessons observed from this large-scale, multi-agency effort identified an information gap among many U.S. military responders in the area of disaster relief operations; this project was born from these efforts. Due to the complex response landscape, as well as the unpredictable nature of disasters, it is important that DoD personnel deployed to support foreign disaster relief efforts are well-equipped to respond to these situations. This handbook is a useful tool to augment service members' knowledge when deployed to these crisis situations. Operational forces responding to those events are likely to encounter a wide range of conditions. Accordingly, it is impossible to provide a single functional concept of operations that can be applied to an extremely broad spectrum of scenarios. Instead, this handbook offers an overarching guide and reference for military responders in disaster relief operations.

i

In addition to serving as a tool for U.S. military personnel, this handbook provides other USG department and agencies and international partners, such as the United Nations, donor countries, regional organizations and non-governmental organizations, with information on DoD's support to DoS and USAID in foreign disaster relief operations and the Department's processes and procedures. This information is essential to a unified and coordinated response.

The *Department of Defense Support to Foreign Disaster Relief (FDR): Handbook for the Joint Task Force Commander and Below* is an important step in DoD's continuing efforts to improve its support for foreign disaster relief operations, of which no two are the same. I recommend its use throughout the Department to guide future responses.

James A. Schear
Deputy Assistant Secretary of Defense
For Partnership Strategy and Stability Operations

ii

ENDORSEMENTS

The following agencies provided written endorsements to the Department of Defense Support to Foreign Disaster Relief (Handbook for JTF Commanders and Below)

- Department of State, Office of Crisis Management Support
- United States Agency for International Development, Office of US Foreign Disaster Assistance
- United States Africa Command
- United States European Command
- United States Northern Command
- United States Pacific Command
- United States Southern Command
- United States Army, Training and Doctrine Command, Peace Keeping and Stability Operations Institute
- United States Navy, Office of the Chief of Naval Operations, International Engagement
- United States Marine Corps, Security Cooperation Education and Training Center
- United States Air Force, Deputy Chief of Staff Logistics, Installation and Mission Support
- Defense Security Cooperation Agency, Humanitarian Assistance, Disaster Relief, and Mine Action Division
- United States Army, South

Copies of these endorsement letters are available upon request. Requests should be sent to:

Test Director, Army, Joint Test Element
ATTN: CSTE-JT
4316 Susquehanna Avenue
Aberdeen Proving Grounds, MD 21005

PREFACE

The 2010 *National Security Strategy* identifies Foreign Disaster Relief (FDR), under the broader mission of humanitarian assistance, as a directed mission of the Department of Defense (DOD). The Foreign Humanitarian Assistance/Disaster Relief Quick Reaction Test was sponsored by United States Southern Command and United States Army South. The project was endorsed by United States Pacific Command, United States Northern Command, the United States Air Force and the Department of Homeland Security, Federal Emergency Management Agency. The Army Test and Evaluation Command served as the Operational Test Agency.

PURPOSE

This handbook provides a Concept of Operations (CONOPS) and Tactics, Techniques, and Procedures (TTP) for joint forces at the operational and tactical levels tasked to perform Foreign Disaster Relief (FDR) operations in support of the Department of State and US Agency for International Development and in coordination with International Organizations such as the United Nations and International Red Cross and Red Crescent movement, other Intergovernmental Organizations (IGO) and Non-Governmental Organizations (NGO). It is intended to be an overarching guide as opposed to a rigid construct.

SCOPE

This handbook addresses DOD's response to natural disasters in *permissive environments*. While it is not intended for use in *complex emergencies*, this information may have utility for United States Government (USG) responses in *complex emergencies*.

The Interagency Standing Committee (IASC) of the United Nations and USAID/OFDA defines a *complex emergency* as "A humanitarian crisis in a country, region or society where there is total or considerable breakdown of authority resulting from internal or external conflict and which requires an international response that goes beyond the mandate or capacity of any single agency and/or the ongoing United Nations' country program." Complex emergencies are characterized by:

- Extensive violence or loss of life
- Massive displacements of people
- Widespread damage to societies and economies
- The need for large-scale, multi-faceted humanitarian assistance
- Hindrance or prevention of humanitarian assistance by political and military constraints
- Security risks for humanitarian relief workers in some areas

iv

In addition to complex emergencies, this handbook may also be useful for foreign consequence management, or Chemical, Biological, Radiological, Nuclear, and high-yield Explosives (CBRNE) related incidents.

AUDIENCE

The primary audience for this handbook is operational level commands, defined as the Joint Task Force (JTF) and its supporting tactical level organizations that plan and execute FDR missions in support of an Affected State. Additionally, it may be used by a strategic level Geographic Combatant Commander's and staff to facilitate planning efforts, or other USG agencies, international organizations IGO or NGO that interface with DOD.

ADMINISTRATIVE INFORMATION

The proponent for this manual is Headquarters, US Southern Command, ATTN: J-8, Science Director, 9301 NW 33rd Street, Doral, FL 33172. Send comments, recommendations, and proposed changes to this handbook to the USSOUTHCOM contact information located at http://www.southcom.mil/

ACKNOWLEDGEMENTS

The following agencies and commands provided direction and resources in the development of the Department of Defense Support to Foreign Disaster Relief – Handbook for Joint Task Force Commander's and Below.

Director Operational Test and Evaluation (DOT&E), Office of the Secretary of Defense (OSD), Washington, District of Columbia.

The Joint Program Office (JPO) Suffolk, Virginia.

The Joint Test and Evaluation (JT&E) Office, Alexandria, Virginia.

Headquarters, Army Test and Evaluation Command, Aberdeen Proving Grounds, Maryland.

We acknowledge and thank the following individuals who provided invaluable subject matter expertise in the development of the handbook.

Karen Zareski, Director, Office of Crisis Management Support, Executive Secretariat, US Department of State, Washington, District of Columbia.

Bridget Premont, and Lieutenant Colonel Paul Matier, Office of Crisis Management Support, Executive Secretariat, US Department of State, Washington, District of Columbia.

Stacy Gilbert, Bureau of Population Refugees and Migration, US Department of State, Washington, District of Columbia.

Steve Catlin, Tom Frey, Arielle Giegerich, and Kate Legates, Military Liaison Unit, Office of US Foreign Disaster Assistance, Bureau for Democracy Conflict and Humanitarian Assistance, US Agency for International Development, Washington, District of Columbia.

Stephen Katz, Bureau for Democracy Conflict and Humanitarian Assistance, US Agency for International Development, Washington, District of Columbia.

Lieutenant General, P.K. Keen, Office of the Defense Representative-Pakistan for U.S. Central Command, Pakistan.

Rear Admiral Sinclair M. Harris, Director, Irregular Warfare, US Navy OPNAV, Pentagon, Washington, District of Columbia.

Rear Admiral Victor Guillory, Commander, Navy South, 4th Fleet, Mayport, Florida.

Ngoc Clark, Melissa Hanlon, and Anne Knight, Assistant Secretary of Defense, Special Operations Low Intensity Conflict, Partnership Strategy and Stability Operations, Pentagon, Washington, District of Columbia.

Nicola Gurwith, Overseas Humanitarian, Disaster Assistance and Civic Aid, Defense Security Cooperation Agency, Arlington, Virginia.

Paul Chlebo Jr. and John Holloway, Office of the Assistant Secretary of Defense, Networks Information and Integration/Chief Information Officer, Arlington, Virginia.

Diana Parzik, Assistant Secretary of Defense, Health Affairs, Pentagon, Washington, District of Columbia.

Commander Bruno Himmler, Office of the Assistant Secretary for Preparedness and Response, Department of Health and Human Services, Washington, District of Columbia.

Colonel Roberto Nang, Peace Keeping and Stability Operations Institute, US Army Training and Doctrine Command, Carlisle Barracks, Pennsylvania.

Clarisa Lamar, Medical Doctrine Literature Division, Directorate of Combat and Doctrine Development, Army Training and Doctrine Command, Fort Sam Houston, Texas.

Mr. Michael D. Burke, Combined Arms Doctrine Directorate, Combined Arms Center, Army Training and Doctrine Command, Fort Leavenworth, Kansas.

Lieutenant Colonel Valeri Jackson, Lieutenant Colonel Anthony P. Terlizzi, and Master Sergeant Theodore Yntema, Security Cooperation Education and Training Center, United States Marine Corps, Quantico, Virginia.

Mark A. Henning, Deputy Director, Navy Lessons Learned System, Naval Warfare Development Command, Norfolk, Virginia.

Corey Hamilton, Headquarters, US Africa Command, Kelley Barracks, Stuttgart, Germany.

Colonel John C. Hope, US European Command Liaison Officer and Lieutenant Colonel Jose Figeroa-Seary, J-9, US European Command, Stuttgart, Germany.

Colonel Fred Humphrey and Master Sergeant Gregory Doles, 361st Civil Affairs Brigade, US Army Reserves, US Army Europe, Panzer Kaserne, Kaiserslautern, Germany.

Lieutenant Colonel Timothy Amoroso, J4 Engineer, US Northern Command, Colorado Springs, Colorado.

Lieutenant Colonel Jeffrey Davis and Lieutenant Commander Derek Beatty, J-55, US Northern Command, Colorado Springs, Colorado.

Lieutenant Colonel Christopher R. Brown, Staff Judge Advocate, US Northern Command, Colorado Springs, Colorado.

Joe Uson, and Major J.J. Garcia, Headquarters, J3, US Pacific Command, Camp Smith, Hawaii.

Commander Joyce B. Blanchard, Commander Pacific Fleet, Honolulu, Hawaii.

Brigadier General Gerald W. Ketchum, Director of Theater Engagement, J-7, US Southern Command, Doral, Florida.

Juan A. Hurtado, Science Director, J-8 Directorate, US Southern Command, Doral, Florida.

COL Mark Walters, Commander, Standing Joint Force Headquarters, US Southern Command, Doral Florida.

Lieutenant Colonel Kevin Rants, Deputy Commander, Standing Joint Force Headquarters, US Southern Command, Doral Florida.

Colonel, Eric A. Snadecki, Director of Plans and Exercises, J-7 US Southern Command, Doral, Florida.

Lieutenant Colonel Patrick J. Wells, Chief, Operations Law US Southern Command, Doral Florida.

Lieutenant Colonel Vincent E. Grizio and Mr. Ricky Stuart, J-8, US Southern Command, Doral Florida.

Colonel Janice Higuera, Deputy Chief of Staff, Engineer, United States Army, South, San Antonio, Texas.

Gregory S. Parker, N-1, Commander Timothy Veschio and Lieutenant Commander Denise Judge, N5, US Navy South, Mayport, Florida

Phil Kearley, and Ronald Reichelderfer, US Joint Forces Command, Suffolk, Virginia.

Master Sergeant Christopher Gay, Lieutenant Colonel Frederick Little, Robert Rackley, and John R. Armstrong, US Special Operations Command, MacDill Air Force Base, Florida.

Lieutenant Colonel Basil R. Cantanzaro, 95th Civil Affairs, Fort Bragg, North Carolina.

Jeff Ackerson, J-5, Joint Task Force Port Opening US Transportation Command, Scott Air Force Base, Illinois.

The 597th Transportation Brigade, and 688th, 689th and 690th Rapid Port Opening Elements, Fort Eustis, Virginia.

Tom Dolan, Bobby Ray Gordon, Victoria Hart, and Jim Welch Center for Excellence, Disaster Management Humanitarian Assistance, Tripler Army Medical Center, Hawaii.

Albert M. Lidy, Institute for Defense Analysis, Alexandria, Virginia.

Michael Dziedzic, US Institute for Peace, Washington, District of Columbia.

William J. White, Jeffrey Petersen, Jamie Weikle, Edward.A. McCulloch, Edward J. Cowan and P.D. Marghella, Contractor support.

Lieutenant Colonel Kimberly A. Enderle, Project Manager and Captain Edwin L. Kolen, Deputy Project Manager.

viii

CHAPTER REFERENCE GUIDE

Chapter 1—United States Government FDR Roles and Responsibilities

Chapter 2—United States Government Foreign Disaster Relief Process

Chapter 3—International Participants and Guiding Principles

Chapter 4—Coordination, Collaboration, Cooperation, Comms, & Cultural

Chapter 5—Operational Context and Mission Planning Considerations

Chapter 6—Disaster Typology

Chapter 7—Planning and Execution

Chapter 8—Safety

Chapter 9—Regional Response Organizations

Chapter 10—Department Of Defense Cross-Cutting Organizations

Chapter 11—DOD Tactical Response Organizations in FDR Operations

Appendix A—Legal Aspects of Foreign Disaster Relief

Appendix B—DOD Guidelines for Interaction with NGOs

Appendix C—Report Formats

Appendix D—Useful Foreign Disaster Related Websites

Appendix E—Training

Appendix F—References

Appendix G—Acronyms

UNCLASSIFIED

This page intentionally left blank

TABLE OF CONTENTS

UNCLASSIFIED

UNCLASSIFIED

TABLE OF FIGURES

UNCLASSIFIED

NAVIGATING THIS HANDBOOK

This handbook is divided into four major sections—introduction; operational context and planning factors; supported and supporting organizations; and appendices.

Section I: Introduction

Chapter 1 provides the background, legal authorities, and guidance for performing FDR operations. It also describes the roles and responsibilities of the various US Government agencies that may be called upon to participate in foreign disaster relief.

Chapter 2 provides an overview of the US Government's foreign disaster response processes.

Chapter 3 describes the humanitarian principles, international guidelines, roles, and responsibilities of the international entities that participate in FDR operations.

Chapter 4 explains the importance of coordination, collaboration, cooperation, and communication functions as well as cultural awareness in a FDR environment.

Section II: Operational Context and Planning Factors

Chapter 5 provides the operational context in a notional five-phase FDR operation, and discusses assessments and metrics.

Chapter 6 details the characteristics of natural disasters and the types of missions that DOD units may be asked to perform in support of FDR.

Chapter 7 details the functions of a JTF commander, primary staff, and special staff during each phase of a FDR operation.

Chapter 8 details safety, force protection (FP), and force health protection (FHP) considerations, and composite/operational risk management processes.

Section III: Supported and Supporting Organizations

Chapter 9 provides a broad overview of the Geographic Combatant Commands (GCC) similarities and differences.

Chapter 10 describes the functional commands and the cross-cutting organizations roles and missions in support of FDR.

Chapter 11 provides an overview of typical DOD tactical units tasked to support FDR operations.

Section IV: Appendices

Appendix A presents the legal aspects of FDR operations.

Appendix B provides the DOD guidelines for interaction with NGOs in a permissive environment.

Appendix C provides sample formats that are useful to staff elements supporting FDR operations.

Appendix D provides a list of useful websites.

Appendix E provides a list of training courses.

Appendix F lists references useful in planning and executing FDR missions.

Appendix G is a list of acronyms used in this handbook.

Color-coded boxes in this handbook contain important information.

Quotation boxes contain quotes pertinent to the subject area.

Blue boxes provide references.

NOTE: Green boxes provide explanatory information on topics of interest or expanded procedural information.

CAUTION: Yellow boxes provide information on potentially hazardous situations, which could result in injury to personnel or damage to equipment.

WARNING: Red boxes provide information on potentially hazardous situations, which could result in serious injury, loss of life, or destruction of equipment.

CHAPTER 1 - UNITED STATES GOVERNMENT FOREIGN DISASTER RELIEF ROLES AND RESPONSIBILITIES

1.1 Overview

The President and the National Security Staff determine the degree to which United States Government (USG) officials and organizations will be involved in Foreign Disaster Relief (FDR) operations, based on strategic considerations delineated in the following documents:

- The National Security Strategy
- The National Defense Strategy
- The National Military Strategy
- Defense Planning Guidance
- Quadrennial Diplomacy and Development Review
- Quadrennial Defense Review

NOTE: *US law refers to foreign humanitarian assistance and disaster relief missions as Foreign Disaster Assistance (FDA). DODD 5100.46 refers to those missions as Foreign Disaster Relief (FDR) and JP 3-29 refers to them as Foreign Humanitarian Assistance (FHA). For the purposes of this handbook, the Department of Defense (DOD) mission is referred to as FDR.*

Legal authority is required prior to USG agencies participating in FDR missions because public funds may only be expended when expressly authorized by Congress. The principal authority is the Foreign Assistance Act (FAA) of 1961, (Public Law 87-195, as amended; 22 United States Code (USC) 2151 et seq.), which provides the guidance for USG engagement with friendly nations and foreign governments. The FAA directs the Department of State (DOS) to provide policy guidance and supervision of programs created within the FAA and provides guidance regarding DOD support to FDR operations.

The remainder of this chapter discusses roles and responsibilities of USG officials and agencies that typically support FDR.

1.2 The President of the United States/Commander-in-Chief and Principal Advisors

The Constitution vests primary authority and responsibility for the formulation and execution of foreign policy in the President. Executive Order 12966 governs the implementation of §404 (Foreign Disaster Assistance) of Title 10, United States Code (USC), which directs the Secretary of Defense to respond to man-made or natural disasters when the Secretary of Defense determines that such assistance is necessary to prevent loss of lives. The Executive Order directs the Secretary of Defense to provide disaster assistance only:

- At the direction of the President
- With the concurrence of the Secretary of State, or
- In emergency situations in order to save human lives, where there is not sufficient time to seek the prior initial concurrence of the Secretary of State, in which case the Secretary of Defense shall advise, and seek concurrence of, the Secretary of State as soon as practicable thereafter

When providing humanitarian assistance, in accordance with the standing executive order, the Secretary of Defense shall consult with the Administrator of United States Agency for International Development (USAID) in the Administrator's capacity as the President's Special Coordinator for International Disaster Assistance. Figure 1-1 is a diagram of interagency coordination.

> **NOTE:** *In accordance with Section 493 of the Foreign Assistance Act (FAA), the President designates a Special Coordinator for International Disaster Assistance. See Section 1.4.2.1.*

1.3 National Security Council

The National Security Council (NSC) is the principal forum that considers and discusses courses of action regarding national security matters, including activities in foreign countries, and subsequently makes recommendations to the President. The President chairs the National Security Council. Its statutory members, in addition to the President, are the Vice President, the Secretary of State and Secretary of Defense. The Chairman of the Joint Chiefs of Staff is the statutory military advisor to the Council, and the Director of Central Intelligence is the intelligence advisor. The Secretary of the Treasury, the U.S. Representative to the United Nations, the Assistant to the President for National Security Affairs, the Assistant to the President for Economic policy, and the Chief of Staff to the President are invited to all meetings of the Council.

Upon declaration of a disaster by a US Ambassador, the President, Secretary of Defense, and the Secretary of State determine the appropriate USG response. The NSC may then convene an International Development and Humanitarian Assistance NSC Interagency Policy Committee (IPC) meeting to review pertinent information and to recommend specific actions.

The NSC/IPC may include:

- Senior DOS and DOD representatives

- Ambassador of Affected State
- USAID representative
- Heads of other USG agencies

INTERAGENCY COORDINATION FOR FOREIGN HUMANITARIAN ASSISTANCE

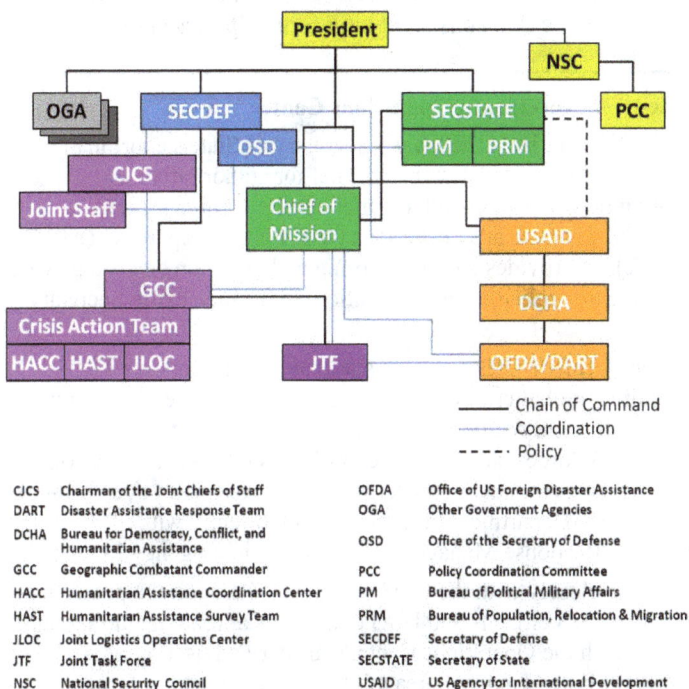

CJCS	Chairman of the Joint Chiefs of Staff	
DART	Disaster Assistance Response Team	
DCHA	Bureau for Democracy, Conflict, and Humanitarian Assistance	
GCC	Geographic Combatant Commander	
HACC	Humanitarian Assistance Coordination Center	
HAST	Humanitarian Assistance Survey Team	
JLOC	Joint Logistics Operations Center	
JTF	Joint Task Force	
NSC	National Security Council	
OFDA	Office of US Foreign Disaster Assistance	
OGA	Other Government Agencies	
OSD	Office of the Secretary of Defense	
PCC	Policy Coordination Committee	
PM	Bureau of Political Military Affairs	
PRM	Bureau of Population, Relocation & Migration	
SECDEF	Secretary of Defense	
SECSTATE	Secretary of State	
USAID	US Agency for International Development	

Figure 1-1: Interagency Coordination for Humanitarian Assistance

1.4 Secretary of State

The Secretary of State is the President's principal foreign policy advisor and is directed by standing Executive Order 12966 to provide or withhold concurrence in the use of armed forces of the United States to provide foreign disaster assistance in response to emergencies.

1.4.1 Department of State

The DOS has many functional responsibilities. This handbook focuses upon the bureaus and offices primarily responsible for FDR operations, which includes:

- Executive Secretariat, Operations Center (S/ES-O)
- Political-Military Affairs (PM)

- Bureau for Population, Refugees, and Migration (PRM)
- Humanitarian Information Unit (HIU)

> **NOTE:** *During international disasters and emergencies, DOS decides if, when, and to what extent, emergency relief is to be provided by the USG consistent with foreign policy goals.*

1.4.1.1 Executive Secretariat, Operations Center

The Executive Secretariat (S/ES) is the Secretary of State's coordination and communications mechanism, and the channel for authoritative communication between DOS and the interagency, foreign affairs community. The Secretariat's Operations Center is the Department's 911 (202-647-1512). It provides 24/7 monitoring and briefs on emerging events as well as secure and reliable communications assistance to the Secretary and senior staff.

The Operations Center's Office of Crisis Management Support (S/ES-O/CMS) monitors global crises and provides crisis contingency planning, including FDR, to DOS regional bureaus and overseas posts. It also supports DOS task forces convened to coordinate embassy draw-downs, ordered evacuations, responses to natural disasters, terrorism events, and other emergencies. Department task forces work directly with USAID/OFDA's Response Management Teams on interagency FDR issues.

CMS also works closely with the DOS Consular Affairs on the protection and evacuation of US citizens and USG employees abroad, including during a disaster, and with the Operations Center's military advisor and the Political Military bureau to keep area military Commanders informed. The Operations Center military advisor also coordinates technical support to the DOS regional bureaus that draft Exec Sec to Exec Sec memorandums requesting DOD assistance in support of FDR operations.

> **NOTE:** *DOS may order an evacuation of US citizens under conditions that involve the use of military assets (although most utilize commercial transportion). Evacuation operations are characterized by uncertainty and may be directed without warning due to sudden changes in a nation's government, reoriented diplomatic or military relationships, a threat to US citizens from internal instability, or a devastating disaster. Some key evacuation planning factors are: situational awareness, appraisal and understanding of the environment, knowledge of constraints and risks.*

1.4.1.2 Political-Military Affairs

The Bureau of Political-Military Affairs (PM) is DOS's principal link to DOD. The PM Bureau provides policy direction in the areas of international security, security assistance, military operations, defense strategy and plans, and defense trade.

Political-Military Affairs advances the DOS-DOD relationship in six key ways:

- Providing the Secretary with a global perspective on political-military issues
- Supporting DOD by negotiating basing agreements, facilitating overseas operations, and by embedding Foreign Policy Advisors with Service branch chiefs and Combatant Commands
- Promoting regional stability by building partnership capacity, strengthening friends and allies through security assistance programs
- Reducing threats from conventional weapons through humanitarian demining and small arms destruction programs
- Contributing to Defense and Political-Military Policy and Planning
- Regulating arms transfers and U.S. defense trade

In addition, PM works closely with the NSS, Geographic Bureaus, Office of the Secretary of Defense (OSD), Joint Staff, and USAID to coordinate crisis management, and complex contingency operations.

1.4.1.3 Bureau for Population, Refugees, and Migration

At the direction of the Under Secretary of State for Democracy and Global Affairs, the Bureau of Population, Refugees, and Migration (PRM) is responsible for formulating policies and coordinating population, refugee, and migration programs, and for administering refugee assistance and US admission programs. PRM promotes its policies through bilateral and multilateral cooperation, and works closely with USAID. PRM administers and monitors US contributions to international and non-governmental organizations to assist and protect refugees abroad.

> **NOTE:** *Under the United Nations Convention Relating to the Status of Refugees of 1951, a refugee is defined as a person who "owing to a well-founded fear of being persecuted for reasons of race, religion, nationality, membership of a particular social group, or political opinion, is outside the country of his or her nationality, and is unable to or, owing to such fear, is unwilling to avail himself of the protection of that country." PRM is the lead federal agency for refugees.*

> **NOTE:** *An Internally Displaced Person (IDP) is someone who is forced to flee his or her home, but who remains within his or her country's borders. IDPs are often incorrectly referred to as refugees. USAID/OFDA is the lead federal agency for dealing with IDPs.*

1.4.1.4 Humanitarian Information Unit

The Humanitarian Information Unit (HIU) is an interagency collaborative effort led by the DOS and the Bureau for Intelligence and Research, Office of Geographic and Global Issues. The purpose of the HIU is to improve US Government planning and response to overseas humanitarian crises and complex emergencies by providing better baseline data, streamlining data requirements, and enhancing information management and dissemination. The core functions of the HIU are:

- Collect, disseminate, and promote the use of unclassified data for overseas humanitarian crises in which USG, civilian, and/or military agencies are directly or indirectly engaged
- Rapidly collate standardized unclassified "core" data sets
- Provide geo-referenced and packaged portable information management tools for US-supported field missions
- Help coordinate USG responses to requests for unclassified information from non-USG agencies and foreign governments during complex emergencies

> **NOTE:** *The HIU collects, interprets, and distributes humanitarian assistance data intended for use by national and field-level decision-makers to assess conditions and factors affecting humanitarian and peacekeeping operations. The HIU does not simply operate as a data repository. Their main function is data analysis.*

1.4.1.5 Geographic Bureaus

DOS is comprised of geographic bureaus, which are responsible for managing the USG's relationships with other countries/governments. An Assistant Secretary of State who reports to the Under Secretary of State for Political Affairs leads each geographic bureau. These geographic bureaus are further comprised of regional bureaus and sub-regional offices that include a director, deputy director, and desk officers for individual countries. The geographic bureaus work closely with US embassies and consulates overseas, and with foreign embassies in Washington, D.C. DOS includes the following geographic bureaus: African Affairs (AF), East

Asian and Pacific Affairs (EAP), European and Eurasian Affairs (EUR), Near Eastern Affairs (NEA), South and Central Asian Affairs (SCA), and Western Hemisphere Affairs (WHA). The Bureau of International Organization Affairs develops and implements US policy with regard to the United Nations and its affiliated agencies, and with certain other international bodies. Figure 1-2 depicts the Geographic Bureaus.

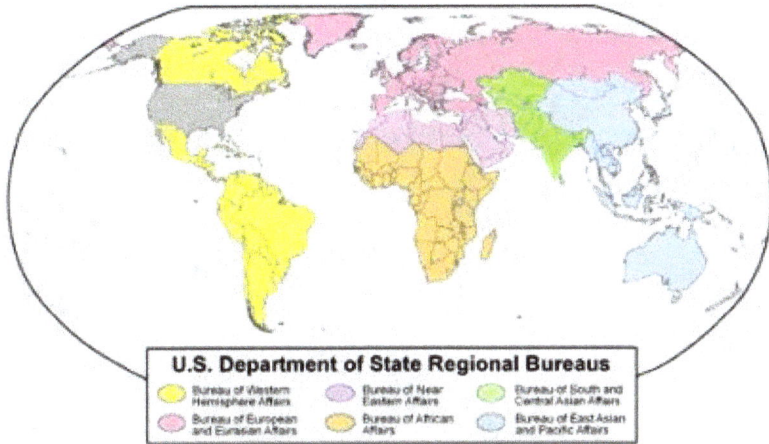

U.S. Department of State Regional Bureaus

Bureau of Western Hemisphere Affairs Bureau of Near Eastern Affairs Bureau of South and Central Asian Affairs

Bureau of European and Eurasian Affairs Bureau of African Affairs Bureau of East Asian and Pacific Affairs

Figure 1-2: DOS Geographic Bureaus

NOTE: *When a foreign disaster is declared, the DOS geographic bureau of the affected area becomes the key participant.*

1.4.1.6 US Diplomatic Mission, Ambassador, Chief of Mission and Country Team

A US diplomatic mission is an integrated structure, led by an Ambassador or a Chief of Mission (COM), who serves as the personal representative of the President. The term, *Chief of Mission* is generally applied to the senior diplomat, including the Ambassador, assigned to head the mission. The terms Ambassador and Chief of Mission are used interchangeably in this handbook. In smaller embassies, the *Charges d'Affaires* will function as the Chief of Mission.

A country team is comprised of the COM, and all agencies represented in the country, such as USAID, Foreign Agricultural Service, Foreign Commercial Service, Peace Corps, military groups, law enforcement, and Defense attachés, as well as the traditional State Department functions concerned with political, economic, consular, management, and public diplomacy issues.

Figure 1-3 shows a typical organizational chart for a US embassy.

Embassy Organization

```
                        ┌──────────────┐
                        │  Ambassador  │
                        └──────────────┘
                               │
                        ┌──────────────┐
                        │ Deputy Chief │
                        │  of Mission  │
                        └──────────────┘
                               │
  ┌──────┬──────────┬──────────┬─────────┬──────────────┬────────┬───────┐
┌──────┐┌────────┐┌────────┐┌────────┐┌──────────┐┌────────┐┌───────┐
│Other ││Political││Economic││Consular││ Senior   ││ Mgmt  ││ USAID │
│Agenc.││Section ││Section ││Section ││Defense Rep││Section││       │
└──────┘└────────┘└────────┘└────────┘└──────────┘└───────┘└───────┘
```

| Other Agencies | Political Section | Economic Section | Consular Section | Senior Defense Rep | Mgmt Section | USAID |

| Defense Attaché Office | Security Assistance Office |

| Public Affairs Office | Regional Security Office |

Figure 1-3: US Embassy Organizational Chart

In addition to serving as the President's representative in foreign countries, the COM validates the need for humanitarian assistance or foreign disaster relief based on the USAID/OFDA criteria described in Section 2.2.2.

The COM is assisted in validation of need by the Country Team, which includes, among other important advisors, a *Mission Disaster Relief Officer* (MDRO).

The MDRO is the focal point for disaster-related information, planning, and response activities relevant to the Affected State. The MDRO is appointed by the COM. The MDRO is a member of the post's Emergency Action Committee and is responsible for preparing Appendix J of the *Emergency Action Plan*, entitled *Assistance to Host Country in a Major Accident or Disaster*, usually referred to as the *Mission Disaster Relief Plan*. The MDRO is familiar with Affected State disaster authorities and capabilities and other potential humanitarian partners in-country, and continually liaises with the USAID/OFDA Regional Advisor, DOD staff on the Country Team, and the State Department's Bureau of Population, Refugees, and Migration.

After a disaster has been declared, the MDRO is responsible for drafting the Disaster Declaration Cable and submitting it to the COM for approval; sharing information about the welfare and whereabouts of American citizens in the Affected State with the post's consular section; keeping a log of important events and report cables; and maintaining regular contact with relevant organizations, such as the UN, donor countries, NGOs, IGOs, etc.

NOTE: *In most US Embassies, assignment as the MDRO may be a collateral or additional duty.*

The Alternate MDRO may be a citizen of the Affected State that works at the American Embassy or US Mission, usually referred to as Foreign Service National (FSN), or Locally Engaged Staff (LES). FSN/LES are experts in the Affected State's culture, local government, and ministries, and are well versed in Affected State relief agencies, and societies.

Secretary of State Message, Subject: USAID/DHCA Office of US Foreign Disaster Assistance Guidance for Disaster Planning and Response. This annual cable provides guidance to all posts concerning support from USAID's Democracy, Conflict, and Humanitarian Assistance (DCHA), Office of US Foreign Disaster Assistance (USAID/OFDA), before, during, and after the occurrence of natural disasters abroad.

1.4.2 United States Agency for International Development Administrator

The USAID Administrator formulates and executes US foreign economic and development assistance policies and programs, subject to the foreign policy guidance of the President, Secretary of State, and the National Security Council. Under the direct authority and foreign policy guidance of the Secretary of State, the Administrator serves as the principal advisor to the President and the Secretary of State regarding international development matters. The Administrator oversees appropriations made under the Foreign Assistance Act of 1961 and directs overall Agency activities in the US and abroad. In addition to leading USAID, the Administrator serves as the President's Special Coordinator for International Disaster Assistance.

1.4.2.1 President's Special Coordinator for International Disaster Assistance

By White House Memorandum, *Designation of the US Agency for International Development Administrator as the Special Coordinator for International Disaster Assistance,* dated 15 SEP 93, the President designated the Administrator of USAID as the Special Coordinator for International Disaster Assistance, as authorized by § 493 of the Foreign Assistance Act (FAA) 1961.

1.4.3 United States Agency for International Development

The United States Agency for International Development (USAID) provides economic development and humanitarian assistance around the world in support of the foreign policy goals of the United States. Although

1-9

a separate agency from the Department of State, it shares certain administrative functions with DOS and reports to and receives overall foreign policy guidance from the Secretary of State. USAID plays a major role in US foreign assistance policy and a principal role in interagency coordination.

> **NOTE:** *USAID is the USG's Lead Federal Agency (LFA) for coordinating all aspects of foreign disaster relief.*

1.4.3.1 Bureau for Democracy, Conflict, and Humanitarian Assistance

USAID's Bureau for Democracy, Conflict, and Humanitarian Assistance (DCHA) coordinates USAID's democracy programs, international disaster assistance, emergency and developmental food aid, aid to manage and mitigate conflict, and volunteer programs. Within DCHA, the Office of US Foreign Disaster Assistance (USAID/OFDA), the Military Liaison Unit (MLU) is the key organization that interacts with DOD during FDR operations.

1.4.3.2 Office of US Foreign Disaster Assistance

Within the USAID Bureau for Democracy, Conflict, and Humanitarian Assistance (DCHA), the Office of US Foreign Disaster Assistance (USAID/OFDA) is delegated the responsibility to provide international disaster and humanitarian assistance and coordinate the USG response to declared disasters in foreign countries. USAID/OFDA's mandate is to:

- Save lives
- Alleviate human suffering
- Reduce the economic and social impact of disasters

OFDA provides technical support to the Administrator of USAID, who serves as the President's Special Coordinator for International Disaster Assistance. OFDA formulates US foreign disaster assistance policy in coordination with other government agencies. It coordinates with USAID offices and others to provide relief supplies (e.g., blankets, plastic sheeting, sanitation hygiene kits), funds implementing partners (e.g., UN agencies, NGOs, Red Cross) to provide direct support and humanitarian assistance, and develops and manages logistical, operational, and technical support for disaster responses.

Besides its coordination activities within the USG, USAID/OFDA carries out these response options in coordination with the Affected State, other donor countries, UN, NGOs, and IGOs. USAID/OFDA response capabilities are discussed in Section 2.3.

> **NOTE:** *USAID/OFDA's responsibility and authority are specified in the Foreign Assistance Act of 1961, as amended, § 491-493 and from delegated Presidential Authority.*

Disaster Assistance Response Teams. With the concurrence of the Ambassador, a Disaster Assistance Response Team (DART) may deploy to an Affected State to assist in the coordination of disaster relief efforts. The DART provides an operational presence capable of carrying out sustained response activities that may include the following:

- Providing technical assistance to the US Ambassador in formulating and executing an appropriate USG response to the disaster
- Developing and, on approval, implementing USAID's response strategy
- Continuing to assess and report on the disaster situation and recommend follow-up actions, including suggested funding levels
- Coordinating the movement and consignment of relief commodities
- Analyzing existing capacity of the infrastructure and relief agencies to ensure an appropriate, efficient response
- Reviewing and recommending approval for (or approving, when delegated the authority) relief program proposals
- Assisting in the coordination of the USG's relief efforts with the Affected State, other donors and relief agencies and, when present, other USG entities, including the US military
- Monitoring and evaluating USAID/OFDA-funded relief activities
- Arranges for the deployment of search and rescue teams based in:
    - Fairfax County, Virginia
    - Los Angeles County, California

> **NOTE:** *Fairfax County, Virginia Fire and Rescue Department and Los Angeles County, California Fire and Rescue Department each have cooperative agreements with USAID/OFDA to train, equip, maintain, and be prepared to deploy as needed, one Urban Search and Rescue (USAR) Task Force each as part of a DART to international disasters requiring, USAR capabilities. Members of these teams are trained to work in an international environment, which includes experience with international USAR standards and international USAR teams. Some members deploy individually filling other positions on the DART.*

The DART organizational chart is shown in Figure 1-4.

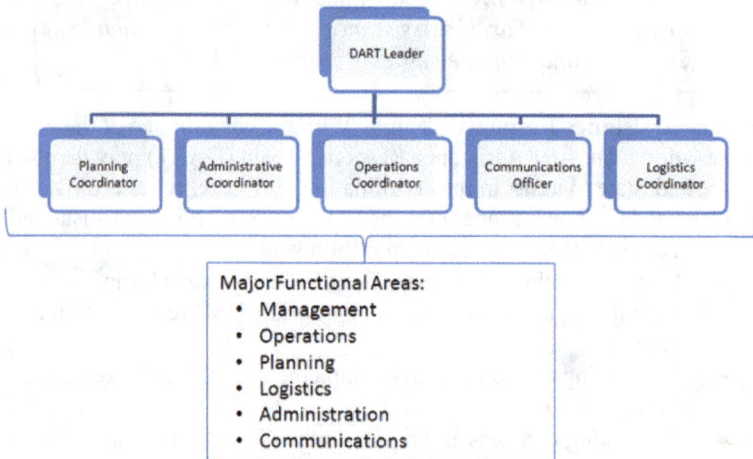

Figure 1-4: DART Organizational Chart

Response Management Team (RMT). If a DART is deployed, an RMT will stand up at USAID headquarters in Washington, DC, which is USAID's equivalent of a Joint Operations Center. The RMT provides support to the DART and coordinates USG disaster response strategy and activities in Washington, DC. The DART relays information on activities, needs, and makes recommendations for appropriate USG assistance to the RMT. The RMT is organized into three major functional areas (management, planning, and operations) and serves as the primary liaison between the DART, other USAID entities, federal agencies, and Congress.

USAID Field Operations Guide (FOG). USAID/OFDA developed the FOG as a reference tool for disaster assessment and response.

The FOG includes:

- Formats and reference material for assessing and reporting on populations-at-risk (PAR)
- DART position descriptions and duty checklists
- Sample tracking and accounting forms
- Descriptions of USAID/OFDA stockpile commodities
- General information related to disaster activities
- Information on working with the military in the field
- A glossary of acronyms and terms used by USAID/OFDA and other organizations with which USAID/OFDA works

The FOG is available online at:
http://www.usaid.gov/our_work/humanitarian_assistance/disaster_assistanc
e/resources/pdf/fog_v4.pdf

Contacts at USAID/OFDA

USAID/OFDA maintains regional offices worldwide to assist in responding
to disasters and developing risk management strategies. Regional advisors
are an excellent source of information and advice throughout a disaster
response. Before contacting a USAID/OFDA regional office directly, first
contact the USAID/OFDA Humanitarian Assistance Advisor/Military
(HAA/M) representative at the GCC, who can provide initial assistance and
make contacts with the Regional offices as appropriate. HAA/Ms serve as
foreign disaster relief subject matter experts. Their roles and
responsibilities include:

- Participating in FDR strategy and contingency planning activities,
  training, and exercises
- As an SME assist in developing the FDR aspect of the GCC's
  strategic plans and crisis action planning products
- Participating in the GCC Crisis Action Team
- Recommending appropriate components/capabilities in FDR
  operations
- Providing informational briefings to GCC regarding
  USAID/OFDA activities and engagements
- Participating in After Action Reviews (AAR)
- Deploying as a member of a DART or with GCC forces in FDR
- Advising on established methods of support to FDR as well as
  humanitarian principles and best practices
- Coordinating with other USAID staff, as well as members of the
  Joint Inter-Agency Coordination Group
- Coordinating with the humanitarian relief community

USAID/OFDA has Memoranda of Understanding/Agreements with various
government agencies that are capable of making contributions to FDR
operations. Those organizations are described in Section 1.6.

> **NOTE:** *OFDA's response to FDR does not include food items;
> that responsibility is the purview of the USAID, Office of Food
> for Peace.*

1.4.3.3 Office of Food for Peace (Food for Peace Program)

General relief objectives of the USAID Food for Peace Program (FPP)
include providing food to low-income and other vulnerable individuals and
populations who are unable to meet basic needs for survival and human

dignity. Individuals may be unable to meet these needs due to an external shock, such as a natural disaster or war, or due to socioeconomic circumstances, such as age, illness, disability, or discrimination. Such individuals are often dependent to some extent upon outside resources to meet their basic food and livelihood needs. Activities include provision of general or supplementary on-site or take-home rations through unconditional safety nets, and food support to institutions assisting the destitute, terminally ill, or highly vulnerable children and youth.

1.4.3.4 Office of Military Affairs

Office of Military Affairs (OMA) plays an important role in USAID interaction with DOD across the full spectrum of stability operations. OMA addresses areas of common interests between defense and development, with a focus on improving civilian-military field readiness programs and coordination. OMA serves as the agency-wide unit for managing the day-to-day aspects of the USAID-military relationship. OMA assigns senior development officers to each GCC and receives LNOs in return.

1.5 Secretary of Defense

The Secretary of Defense (SECDEF) is the principal defense policy advisor to the President and is responsible for the formulation and execution of defense policy. Under the direction of the President, the SECDEF exercises authority, direction, and control over the DOD. The SECDEF is a member of the President's Cabinet and of the National Security Staff.

Under normal conditions, working under the guidance of the President, and with the concurrence of the Secretary of State, SECDEF directs DOD support to USG humanitarian operations foreign disaster assistance and establishes appropriate command relationships.

The SECDEF has delegated the provision to provide assistance to save lives in emergencies under Executive Order 12966 to the Geographic Combatant Commanders.

1.5.1 Office of the Secretary of Defense

The Office of the Secretary of Defense (OSD) is the principal staff element of the Secretary in the exercise of policy development, resource management, fiscal, and program evaluation responsibilities. The following organizations within OSD are responsible for developing various aspects of policy related to Foreign Disaster Relief:

- Under Secretary of Defense for Policy (USDP)
- Assistant Secretary of Defense for International Security Affairs (ASD/ISA)
- Assistant Secretary of Defense for Asia/Pacific Security Affairs (ASD/APSA)

1-14

- Assistant Secretary of Defense for Special Operations and Low Intensity Conflict (ASD/SOLIC)
- Assistant Secretary of Defense for Health Affairs (ASD/HA)
- Assistant Secretary of Defense for Homeland Defense and America's Security Affairs (ASD/HD&ASA)
- Assistant Secretary of Defense for Networks Information and Integration (ASD/NII)

1.5.2 Department of Defense

In addition to its traditional national defense roles, FDR is also a Department of Defense (DOD) directed mission. US military forces have long played a significant role in supporting foreign disaster relief operations. DOD assets may only be used in support of foreign disaster relief operations when the following three criteria have been met:

- The military provides a unique service
- International civilian capacity is overwhelmed
- Civilian authorities request or are willing to accept assistance

> **NOTE:** For the benefit of individuals not familiar with DOD; Chapter 9 discusses Geographic Combatant Commands; Chapter 10 describes the cross-cutting agencies and organizations; and Chapter 11 discusses DOD's unique tactical-level capabilities.

1.5.2.1 Department of Defense Authority

DOD policy dictates that when the US military becomes involved in foreign disaster relief operations:

- The military mission should be clearly defined
- The risks should be minimal
- Other core DOD missions should not be affected

The following directives and instructions provide the authority for DOD participation in FDR operations:

Secretary of Defense Message, May 04, Subject: Policy and Procedures for Department of Defense Participation in Foreign Disaster Relief/Emergency Response Operations. This message from the Office of the Assistant Secretary of Defense for Special Operations and Low Intensity Conflict (ASD/SOLIC) and the Defense Security Cooperation Agency (DSCA) provides guidance on response by regional combatant commands to natural or man-made disasters in their areas of responsibility and elaborates guidance previously approved by the SECDEF.

Department of Defense Directive (DODD) 5100.46, Foreign Disaster Relief. This directive establishes policy guidance for Foreign Disaster Relief operations. It defines FDR as prompt aid used to alleviate the

1-15

suffering of foreign disaster victims. It also provides for Service component participation in FDR operations only after DOS makes a determination that FDR shall be provided.

> **NOTE:** *This policy does not prevent a local military commander, at the immediate scene of a foreign disaster, from undertaking prompt relief operations in response to requests from the local Affected State authorities, when time is of the essence and when, in the estimate of the commander, humanitarian considerations make it advisable to do so. All support requires COM knowledge and approval. This authority is referred to as the 72 Hour Life and Limb provision or immediate response authority and is discussed in Chapter 2.*

Department of Defense Instruction (DODI) 3000.05, Stability Operations. This instruction contains DOD stability operations policy and assigns responsibility for the identification and development of US military capabilities to support stability operations. This instruction establishes stability operations as a core military mission. It outlines DOD policy and assigns responsibility for planning and executing stability operations.

Department of Defense Instruction 6000.16, Military Health Support for Stability Operations. This instruction describes the medical component of stability operations.

1.5.2.2 Chairman of the Joint Chiefs of Staff

The Chairman of the Joint Chiefs of Staff (CJCS) is the principal military advisor to the President and the Secretary of Defense. The Chairman transmits communications to the Geographic Combatant Commanders (GCC) from the President and the SECDEF, but does not exercise military command over operational forces. The CJCS is responsible for recommending military capabilities and appropriate supported and supporting relationships for FDR operations to the SECDEF.

1.5.2.3 The Joint Staff

The Joint Staff (JS) supports the CJCS in meeting responsibilities for the unified strategic direction of the combatant forces, their operation under unified commands, and their integration into an efficient team of land, naval, and air forces.

Primary Joint Staff directorates provide management of joint operational processes, activities, and articulate capabilities associated with the basic joint functions.

Figure 1-5 shows the Joint Directorates and their functions.

Joint Directorates	Function
J-1	Manpower and Personnel
J-2	Intelligence
J-3	Operations
J-4	Logistics
J-5	Strategic Plans and Policy
J-6	Command, Control, Communications, and Computer Systems
J-7	Operational Plans and Joint Force Development
J-8	Force Structure, Resources, and Assessment

Figure 1-5: Joint Directorates

The J-3, J-4, J-5, and J-8 play key roles in validation of FDR requirements.

The Operations Directorate (J-3) assists the Chairman in carrying out his responsibilities and is involved in every aspect of the planning, deployment, execution, and redeployment of US operational forces in response to worldwide crises. The J-3 operates the National Military Command Center (NMCC), which integrates operations, intelligence, and logistics expertise. The J-3 coordinates with OSD to evaluate DOD capabilities, requested by the DOS, in support of FDR. Authorization for a GCC to continue FDR operations, past the initial 72 hours following a foreign disaster, deployment of forces, and to expend Overseas Humanitarian Disaster and Civic Aid (OHDACA) funds, must be authorized by the SECDEF through a CJCS Execution Order.

The Logistics Directorate (J-4) coordinates the logistics and transportation requirements through the Joint Logistics Operations Center (JLOC), located within the NMCC.

The Strategic Plans and Policy Directorate (J-5) coordinates policy, is the principal representative of CJCS in interagency forums related to FDR, and has primary responsibility for the contingency plans (CONPLANs) in support of FDR.

The Force Structure, Resources, and Assessment Directorate (J-8) coordinates with the supported Combatant Commander and OSD (Comptroller) to secure FDR funding.

1.5.2.4 Joint Doctrine

The following publications provide Joint doctrinal guidance related to the FDR mission. This guidance helps to define DOD roles and responsibilities in this mission area.

Joint Publication 3-29, Foreign Humanitarian Assistance. JP 3-29 contains joint doctrine for FDR operations. It provides guidance for the various types of missions in which military assets may be involved. It also discusses interagency coordination, roles and responsibilities, and the principal organizations involved (military, governmental, public, and private sector). Finally, JP 3-29 details the military planning process for FDR as well as the execution and assessment of important aspects of operations.

Joint Publication 3-33, Joint Task Force Headquarters. JP 3-33 provides joint doctrine for the formation and employment of a Joint Task Force Headquarters (JTF-HQ) to command and control joint operations. It provides guidance on the JTF-HQ role in planning, preparing, executing, and assessing JTF operations.

1.5.2.5 Geographic Combatant Commanders

Geographic Combatant Commanders (GCC) direct military operations within their areas of responsibility (AORs). FDR missions are undertaken in response to requirements identified by the Chief of Mission in disaster-affected states. In response to a foreign disaster, the supported GCC typically forms a Joint Task Force (JTF). If the requested requirements are beyond the scope of the forces allocated to the GCC, the GCC will submit a Request for Forces (RFF) to the Joint Staff J-3. Additional information regarding GCCs is included in Chapter 9.

1.5.2.6 Functional Combatant Commands

The following Unified Commands have cross-cutting capabilities for the provision of personnel and resources in support of the geographic combatant commands:

- US Special Operations Command
- US Strategic Command
- US Transportation Command

Additional information can be found in Chapter 10.

1.5.2.7 Defense Agencies

Defense agencies provide support to OSD and conduct programming in specialized areas of concern according to defense policy. Defense agencies that frequently support FDR operations include:

- Defense Information Systems Agency
- Defense Logistics Agency
- Defense Security Cooperation Agency
- National Geospatial Agency

For additional information, refer to Chapter 10.

1.6 Other Government Agencies

The scope of some disasters may require resources from the following USG agencies and departments:

1.6.1 US Department of Agriculture

The mission of the US Department of Agriculture (USDA) is to provide leadership on matters related to agriculture, food safety, natural resources, and related issues. USDA's Foreign Market Development Program works with USAID to assist foreign countries in the development, maintenance, and implementation of agricultural commodity standards, market news reporting services, and livestock and meat grading services. USDA includes the US Forest Service, which may have a direct role in FDR operations.

1.6.1.1 United States Forest Service

The Forest Service's Fire and Aviation Management organization works to advance technologies in fire management; maintain and improve mobilization and tracking systems; and supports federal, state, and international partners in matters related to fire suppression.

The International Programs Office coordinates the Forest Service's international work. The Disaster Assistance Support Program (DASP), within the International Programs Office, provides USAID/OFDA with essential technical support in disaster response management, planning, operations, preparedness, and prevention. DASP can also access capabilities within the Forest Service's Fire and Aviation Management organization to support USAID/OFDA activities.

1.6.2 US Department of Commerce/National Oceanic and Atmospheric Administration

The National Oceanic and Atmospheric Administration (NOAA) is a subordinate organization within the US Department of Commerce and functions as a supplier of environmental information products pertaining to the state of the oceans and the atmosphere through the National Weather Service. Specifically, NOAA seeks to understand and predict changes in climate, weather, oceans, and coasts; to share that knowledge and information with others who require this information for strategic decision-making; and to conserve and manage coastal and marine ecosystems and resources.

1.6.3 US Department of Energy/Nuclear Regulatory Commission

The mission of the Department of Energy is to ensure America's security and prosperity by addressing its energy, environmental, and nuclear challenges through transformative science and technology solutions.

A predictable cascade to large-scale international disasters is damage to national power infrastructure systems and nuclear reactors. The Department

1-19

of Energy, and its subordinate agency the Nuclear Regulatory Commission, play key roles in advising Affected States in energy infrastructure restoration and in controlling damage that has occurred to nuclear power infrastructure.

> **NOTE:** *The 2011 earthquake and tsunami in Japan gave a clear demonstration of the phenomena of cascading effects in disasters and the subsequent damage to critical infrastructure, including nuclear power plants.*

1.6.4 US Department of Health and Human Services

Health and Human Services (HHS) is the USG's principal agency for protecting the health of all Americans and providing essential human services, especially for those who are least able to help themselves.

During an international disaster response, Health and Human Services (HHS) will support USAID/OFDA in cooperation with other USG agencies and international partners. HHS may assist, as requested, at the strategic, operational, and tactical levels. Depending on the nature of the request, HHS personnel may provide technical assistance or implement the response in a supporting role.

HHS is authorized to provide leadership in international programs, initiatives, and policies that deal with public health and medical emergency preparedness and response through the office of the Assistant Secretary for Preparedness and Response (ASPR).

1.6.4.1 Centers for Disease Control and Prevention (CDC)

Working with US states and other partners, CDC provides a system of health surveillance to monitor and prevent disease outbreaks (including bioterrorism), implement disease prevention strategies, and maintain national health statistics. The CDC also provides for immunization services, workplace safety, and environmental disease prevention. Among the most important components with international responsibility is the Coordinating Office for Global Health, which provides national leadership, coordination, and support for CDC's global health partners.

1.6.5 US Department of the Interior/United States Geological Survey

The Department of the Interior/US Geological Survey (USGS) supports FDR in two ways. First, with the Volcano Disaster Assistance Program (VDAP). VDAP teams can be quickly mobilized and deployed worldwide (in support of USAID/OFDA) to assess and monitor hazards associated with volcanoes threatening to erupt. Second, the USGS Earthquake Hazards Program produces *ShakeMap* in conjunction with regional seismic

1-20

network operators. *ShakeMap* sites provide near-real-time maps of ground motion and shaking intensity following significant earthquakes. Those maps are used for post-earthquake response and recovery, public and scientific information, as well as for preparedness exercises and disaster planning.

1.6.6 US Department of the Treasury

The Office of Foreign Assets Control (OFAC) of the US Department of the Treasury administers and enforces economic and trade sanctions based on US foreign policy and national security goals. There are instances when a natural disaster occurs in which the US will provide assistance to a country under sanction. In those cases, when a humanitarian organization such as an NGO is funded to carry out USAID programs in a country where sanctions exist, USAID will apply for an OFAC license through the Department of State. Additionally, the NGO may be required to seek a commercial export license from the Department of Commerce when transferring funds or assets to a country under sanction.

This page intentionally left blank

CHAPTER 2 - US GOVERNMENT FOREIGN DISASTER RELIEF PROCESS

2.1 Overview

The United States Government (USG) response to foreign disasters is not a linear process; it involves nearly simultaneous activities of several key officials and agencies within the Department of State (DOS), US Agency for International Development (USAID), and the Department of Defense (DOD). This chapter describes the process normally used in response to a request for assistance from an Affected State following a natural disaster. Figure 2-1 reflects the USG interagency coordination process that results in DOD participation.

Interagency Coordination flow when USAID/OFDA and DOD respond

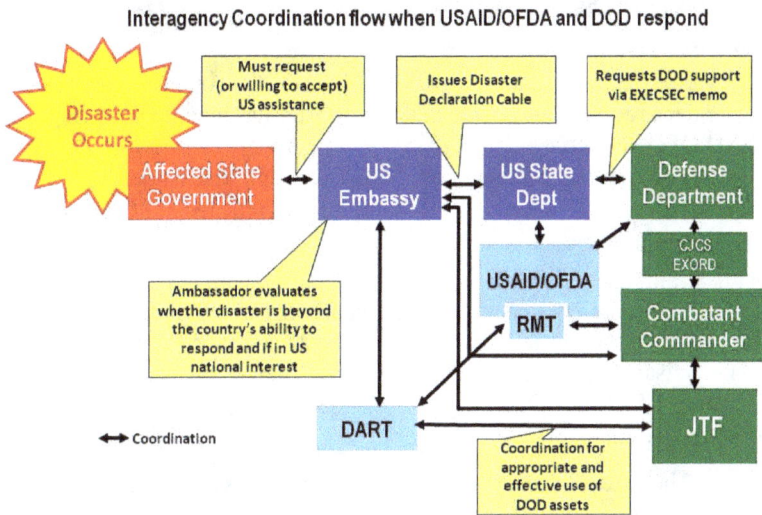

Figure 2-1: Interagency Coordination after a Disaster Occurs

2.2 Department of State and United States Agency for International Development Disaster Response Process

This section describes the process used by the DOS, diplomatic personnel, and USAID in response to foreign disasters.

2.2.1 Mission Disaster Relief Officer

The Mission Disaster Relief Officer (MDRO), described in Section 1.4.1.5, will take several actions when a foreign disaster occurs:

- Verifies the magnitude of the disaster and the impact on the population through Affected State government officials, other embassies, non-governmental organizations (NGOs), the UN,

Intergovernmental Organization (IGO) representatives, and Red Cross and Red Crescent societies

- Notifies the Ambassador and provides an overview of the situation (in some cases, depending upon the scope of the disaster, the mission's emergency action committee will be convened)
- Contacts USAID/OFDA's Regional Advisor in the affected region to ensure communication flow and coordination
- Drafts the Disaster Declaration Cable for approval by the Ambassador, and alerts USAID/OFDA Washington (providing background and current situation information regarding the disaster and the mission's anticipated course of action—sometimes using a Disaster Alert Cable)

2.2.2 Disaster Declaration Cable

When requesting assistance from USAID/OFDA, the Ambassador issues a disaster declaration cable stating that the disaster has met the following USAID/OFDA criteria:

- The Affected State requests, or will accept, assistance
- The disaster is beyond the ability of the Affected State to respond
- Responding is in the interest of the USG

That determination is usually made in consultation with USAID/OFDA Regional Advisors. For countries without an official US diplomatic presence, the Assistant Secretary of State for the appropriate region may declare a disaster via a memorandum from the State Department to the Director of USAID/OFDA.

The cable may include:

- The extent to which the Affected State needs assistance to respond adequately to the disaster
- The intended use of requested resources, including recommended organization(s) through which funds will be channeled
- Estimated size of the population-at-risk
- Estimated number of people injured or killed
- Estimated number of displaced and homeless
- Immediate humanitarian needs
- Disaster background information such as geographic location and damage to infrastructure, crops, and livestock
- Other donor efforts and contributions
- Additional information from assessment reports

2.3 USAID/OFDA Disaster Response Options

The USAID/OFDA disaster response includes several options:

- Provide an initial $50,000 for immediate disaster relief
- Deploy Regional Advisors to the Affected State
- Deploy an Assessment Team
- Provide USG relief commodities from USAID/OFDA warehouses
- Deploy a Disaster Assistance Response Team (DART)
- Stand up a Response Management Team in Washington
- Provide grants/funding to UN, NGOs, and IGOs

If requirements cannot be met using these options alone, USAID/OFDA may request assistance from DOD and other USG agencies.

2.3.1 Provide Initial $50,000

The Ambassador can request up to $50,000 for immediate disaster relief from USAID/OFDA (referred to as "Ambassador's Authority"). Those funds are to be used for immediate disaster relief or rehabilitation and not for long-term reconstruction or to purchase food. Humanitarian and relief assistance are generally designated for organizations involved in initial response rather than for Affected State institutions. Any assistance requested above that amount must be coordinated with, and approved by, USAID/OFDA.

> **NOTE:** *The decision to use additional response options, beyond the initial $50,000, is made by the USAID/OFDA Director based on the magnitude of the disaster and the Affected State's own response capabilities. Except for the deployment of Regional Advisors and Assessment Teams, all USAID/OFDA response options require issuance of a disaster declaration.*

2.3.2 Deploy Regional Advisors

USAID/OFDA Regional Advisors conduct vulnerability and damage assessments; coordinate with other donors; work closely with UN agencies, Red Cross/Red Crescent societies, and NGOs. Regional Advisors also determine need for relief commodities and coordinate with the Country Team on overall USG relief efforts.

2.3.3 Deploy Assessment Team

USAID/OFDA's assessment teams are typically composed of both regional and sector-specific specialists (such as experts in medical and public health, water and sanitation, food safety and nutrition, shelter and mass care, hazard-specific expertise, and logistics), as well as management staff

2-3

familiar with USAID/OFDA policies, procedures, general coordination, and programmatic functions. Assessment teams provide information and recommendations to all concerned to enable timely decisions regarding USG response.

2.3.4 Provide USAID/OFDA Relief Commodities

USAID/OFDA may provide disaster relief commodities (such as blankets, plastic sheeting, and water containers) from their worldwide stockpiles (Miami, Florida; Pisa, Italy; and, Dubai, United Arab Emirates). USAID/OFDA contracts transportation services via sealift or land transport and may also fund air transport of emergency commodities when urgent delivery is required.

> **NOTE:** *USAID/OFDA may request DOD assistance with the transportation of emergency relief commodities when, for example, commercial alternatives are unavailable or when unique military capabilities can expedite relief efforts during urgent, life-saving situations.*

2.3.5 Deploy Disaster Assistance Response Team

USAID/OFDA may deploy a DART into the disaster area to assist in coordination of the FDR effort. The DART assists the Embassy with the management of the USG response to the disaster. The DART leader reports to the Ambassador; ensures USG disaster relief efforts are coordinated; and, concurrently reports to the USAID/OFDA Response Director in Washington to ensure that USAID/OFDA's mandate and mission are carried out effectively. Prior to deployment, the DART leader receives the objectives, priorities, constraints, and reporting requirements for the DART.

DARTs coordinate their activities with the Affected State, NGOs, IGOs, UN relief agencies, and other donor nations. When US military assets are involved with the disaster response, the DART works closely with those assets to ensure a coordinated USG effort.

2.3.6 Stand up a Response Management Team

If a DART is deployed by USAID/OFDA, a Response Management Team (RMT) will stand up at USAID headquarters in Washington, DC, which is USAID's "Joint Operations Center." The RMT provides support to the DART and coordinates USG disaster response strategy and activities in Washington, DC. The DART relays information on activities, needs, and recommendations to the RMT. The RMT is organized into three major functional areas (management, planning, and operations) and serves as the primary liaison between the DART, other USAID entities, federal agencies, and Congress. The RMT takes the lead role in the operational aspects of the disaster response, determining the best method to activate and coordinate resources, including funding, staffing, and relief supplies.

2.3.7 Provide Grants/Funding to UN, NGOs and IGOs

USAID/OFDA can provide funding to implement emergency programs to NGOs, IGOs, and IOs, including UN agencies. NGOs do not have to be US-based, nor do they have to be registered as private voluntary organizations with USAID to be eligible to receive funding. USAID/OFDA works with the MDRO to learn the capabilities within the NGO community before making funding decisions.

> **NOTE:** *The response options listed above can be delivered separately or in combination. However, USAID/OFDA may also choose not to provide assistance for the disaster declaration. If no assistance is provided, the cable to the US Embassy in the Affected State should detail the basis for that decision.*

2.3.8 DOD Participation in FDR

After the Disaster Declaration Cable is transmitted, USAID/OFDA and DOS commence determination of the level of USG assistance. If USAID/OFDA identifies a requirement within an Affected State that can be most effectively met by DOD, DOS will initiate a request through the Office of the Secretary of Defense.

2.4 DOD FDR Response Process

Once specific requirements for DOD's unique capabilities have been identified by USAID/OFDA, they then forward a draft request for assistance to the DOS Executive Secretary (EXECSEC). The DOS EXECSEC reviews the request, and then forwards a memorandum to the DOD EXECSEC specifying the type of assistance requested and whether the support requested is reimbursable or non-reimbursable basis. That memorandum:

- Preserves visibility and decision-making authority for OSD on the use of DOD assets and personnel
- Helps to ensure that any request for assistance is vetted and validated through senior leadership at USAID/OFDA and DOS
- Allows officials in OSD to review the request against other potential demands upon DOD resources

2-5

If the mission is to be undertaken on a non-reimbursable basis, OSD will vet the request with relevant offices and determine, through the Defense Security Cooperation Agency (DSCA), if sufficient resources are available.

See Figures 2-2 and 2-3 for sample memoranda reflecting DOS requests for reimbursable and non-reimbursable support.

Upon approval by the Secretary of Defense or the Deputy Secretary of Defense, the GCC is authorized to commence relief operations. The Joint Staff will then issue an Execute Order (EXORD) to guide the GCC in its response.

DOD and USAID/OFDA may exchange Liaison Officers at all levels to facilitate a coordinated response. The Liaison Officer's mission is to distribute assessments, convey resource requirements, articulate current operations status, and assist with future plans.

> *NOTE: DOD always functions in a supporting role during FDR operations. USAID/OFDA provides coordination and advisory interface between the military and the humanitarian relief community at the strategic level in Washington, DC; at the operational level at the GCC; and at the tactical level with the JTF.*

Consistent with their Title 10 authorities, GCCs and their subordinate commanders, geographically located in an area affected by a natural disaster, and to which the US Ambassador has issued a disaster declaration, are authorized to provide timely, life-saving assistance for up to 72 hours if requested by local authorities (this is commonly referred to as the *immediate response authority*, and is also known as the *72 hour, life and limb* provision). If this provision is utilized the Secretary of Defense (SECDEF) will be notified as soon as practicable. After 72 hours, Commanders should evaluate whether mission support continues to meet the intent of immediate response authority and they must have SECDEF approval to continue support.

USAID
FROM THE AMERICAN PEOPLE

March 11, 2011

UNCLASSIFIED

MEMORANDUM FOR MICHAEL L. BRUHN
EXECUTIVE SECRETARY
DEPARTMENT OF DEFENSE

SUBJECT: Department of Defense Support to the U.S. Agency for International Development for the Japan Earthquake Disaster Response

The U.S. Agency for International Development (USAID) seeks your approval for Department of Defense (DOD) assistance to provide transportation support on a reimbursable basis to the overall U.S. Government relief effort in Japan in response to the 8.9 magnitude earthquake that struck on March 11.

The transportation support would consist of airlift of Non-DoD personnel and equipment from the United States to Japan.

With your approval, USAID will send to DOD for countersignature an executed Reimbursement Agreement, which will contain all necessary financial information.

DOD may contact the USAID/OFDA Financial Team, ▮▮▮▮▮▮▮▮ ▮▮▮▮▮▮ for any concerns regarding this request.

Christa White
Executive Secretary

CC: Executive Secretary, Department of State

U.S. Agency for International Development
1300 Pennsylvania Avenue, NW
Washington, DC 20523
www.usaid.gov

UNCLASSIFIED

Figure 2-2: Example Reimbursable EXECSEC Memo

UNCLASSIFIED

ORIG SENT TO DOD VIA COURIER; COPY TO IPS
ELECTRONIC DIST:
S -
D-L
D-S
F
P
E
T
M
G
R
C
INR
S/P
PA
SES
SES-S
S/ES-CR
WHA
PM
S/ES-O
L
H
RM
S/CPR
IPS/JMB

3/6/2010
12:35 PM

S/ES 201003714
United States Department of State

Washington, D.C. 20520

MAR -5 2010

UNCLASSIFIED

MEMORANDUM FOR LIEUTENANT COLONEL ALFREDO NAJERA, USA
ACTING EXECUTIVE SECRETARY
DEPARTMENT OF DEFENSE

SUBJECT: C-130s for Chile Earthquake Relief Efforts

      The Department of State requests the assistance of the Department of Defense (DoD) in providing humanitarian assistance for disaster relief and recovery surrounding the February 27 earthquake in southern Chile, through the utilization of two C-130s to transport relief commodities. The U.S. Embassy Country Team in Santiago – working closely with USAID's Office of U.S. Foreign Disaster Assistance (OFDA) – will coordinate with the Government of Chile to facilitate this support. On February 28, Ambassador Paul E. Simons declared a disaster in Chile. The Government of Chile has requested and agreed to accept our assistance of relief commodities. This unique capability to rapidly transport large amounts of material to austere locations cannot be reasonably procured in the commercial sector.

      We request DoD assistance be provided on a non-reimbursable basis under such authorities as are available to DoD (e.g., Title 10 U.S. Code 404 Sections and/or 2561) through the use of funds from the Overseas Humanitarian, Disaster, and Civic Aid (OHDACA) account. OFDA concurs with utilization of DoD assets in support of these efforts.

      The Office of the Legal Adviser cleared on this message. The Department of State's point of contact for this request is Chile Desk Officer ███████████████, cell ███████████. The Office of the Secretary of Defense's point of contact is ███████████ in the Office of the Deputy Assistant Secretary for Defense for Partnership Strategy & Stability Operations at ███████████.

Keran Azoradie for
Daniel B. Smith
Executive Secretary

UNCLASSIFIED

Figure 2-3: Example Non-reimbursable EXECSEC Memo

UNCLASSIFIED

2.5 Other US Government Organizations

USAID/OFDA has standing agreements with the following US government entities:

- Department of Agriculture/US Forest Service
- Department of Commerce
    - US Census Bureau
    - National Institute of Standards and Technology
    - National Oceanic and Atmospheric Administration
- Department of Health and Human Services/
    - Agency for Toxic Substance and Disease Registry
    - Centers for Disease Control and Prevention
- Department of Homeland Security
    - US Coast Guard
    - Federal Emergency Management Agency
- Department of Interior/US Geological Survey
- Department of Treasury
- General Services Administration
- National Science Foundation
- Peace Corps

Subsequent to a natural disaster, USAID/OFDA may elect to request support from one of the aforementioned agencies in accordance with previously established agreements or they may elect to establish a new agreement with other agencies and entities where appropriate.

This page intentionally left blank

UNCLASSIFIED

CHAPTER 3 - INTERNATIONAL PARTICIPANTS AND GUIDING PRINCIPLES OF INTERNATIONAL HUMANITARIAN ASSISTANCE

3.1 Overview

This chapter describes the humanitarian principles, international guidelines, and the roles and responsibilities of the international entities that participate in foreign disaster relief (FDR) operations.

3.2 Humanitarian Principles

The guiding principles of *humanity, neutrality, impartiality,* and *operational independence*, adopted by the United Nations (UN), form the basis of international humanitarian assistance.

- *Humanity.* Human suffering must be addressed wherever it is found; the purpose of humanitarian action is to protect life, health, and ensure respect for human beings
- *Neutrality.* Humanitarian participants must not take sides in hostilities or engage in controversies of a political, racial, religious, or ideological nature
- *Impartiality.* Humanitarian action must be carried out on the basis of need alone, giving priority to the most urgent cases of distress and making no distinctions on the basis of nationality, race, gender, religious belief, class, or political opinion
- *Operational Independence.* Humanitarian action must be autonomous from the political, economic, military, or other objectives that any organization may hold with regard to areas where humanitarian action is being implemented

> **NOTE:** *Humanitarian relief should be provided under the general principle of* **Do No Harm**.

Improperly delivered aid can divide communities, fuel conflicts, force displacement, and build cultures of dependency. Consistent with the guiding principles, effective delivery of humanitarian aid will:

- Provide for the needs of the Affected State, with special emphasis on the needs of women, children, and persons most vulnerable
- Be consistent with internationally recognized standards
- Be provided by trained personnel with the involvement of the affected population and in a manner respectful of the local customs, religions, and culture

These principles are the underpinnings that guide actions and help define the roles and responsibilities of participants, whether they are Non-

Governmental Organizations (NGO), Intergovernmental Organizations (IGO), the UN, or military/civil defense forces participating in the response effort. Guidelines for DOD interaction with NGOs are provided in Appendix B.

3.3 International Guidelines

3.3.1 Oslo Guidelines

The Oslo Guidelines, formally titled *Guidelines on the Use of Foreign Military and Civil Defence Assets in Disaster Relief* (rev. 1.1, Nov. 2007), establish the basic framework for formalizing and improving the effectiveness and efficiency of the use of foreign Military and Civil Defense Assets (MCDA) in international disaster relief operations. The UN Inter-Agency Standing Committee (IASC) and the UN humanitarian agencies have agreed to these guidelines. While a significant number of Member States have participated in the development of the guidelines and endorsed their use, they are not binding on Member States. Implementing and operational partners are encouraged to follow this guidance. Member States and regional organizations engaged in relief or military operations in response to natural disasters are also encouraged to use the principles and procedures provided. The Oslo guidelines:

- Address the use of MCDA following natural, technological and environmental emergencies in times of peace
- Cover the use of United Nations (UN) MCDA resources requested by the UN humanitarian agencies and deployed under UN control specifically to support humanitarian activities, as well as, other foreign MCDA that might be available
- Provide principles, concepts, and procedures for requesting and coordinating MCDA when these resources are deemed necessary and appropriate, and for interfacing with foreign military forces who are conducting activities which impact on UN humanitarian activities
- Are primarily intended for use by UN humanitarian agencies and their implementing and operational partners, Resident and Humanitarian Coordinators, MCDA commanders and commanders of other deployed forces performing missions in support of the UN humanitarian agencies and liaison officers coordinating UN humanitarian activities with foreign military forces

The following terms are essential for establishing a common understanding of the terminology used by the Oslo Guidelines.

Humanitarian Assistance. Humanitarian assistance is aid to an affected population that seeks, as its primary purpose, to save lives and alleviate

3-2

suffering of a crisis-affected population. For the purposes of these guidelines, assistance can be divided into three categories based on the degree of contact with the affected population.

- *Direct Assistance* is the face-to-face distribution of goods and services
- *Indirect Assistance* is at least one step removed from the population and involves such activities as transporting relief goods or relief personnel
- *Infrastructure Support* involves providing general services, such as road repair, airspace management and power generation that facilitate relief, but are not necessarily visible to or solely for the benefit of the affected population

These categories are important because they help define which types of humanitarian activities might be appropriate to support with international military resources under different conditions, given that ample consultation has been conducted with all concerned parties to explain the nature and necessity of the assistance.

International Disaster Relief Assistance. In the context of the present Guidelines, International Disaster Relief Assistance (IDRA) means material, personnel, and services provided by the international community to an Affected State to meet the needs of those affected by a disaster. It includes all actions necessary to grant and facilitate movement over the territory, including the territorial waters and the airspace, of a Transit State. IDRA delivered in accordance with the humanitarian principles identified above is humanitarian assistance.

Military and Civil Defense Assets (MCDA). MCDA are comprised of relief personnel, equipment, supplies, and services provided by foreign military and civil defense organizations for IDRA. Further, for the purpose of this handbook, civil defense organization means any organization that, under the control of a Government, performs the functions enumerated in paragraph 61 of Additional Protocol I to the Geneva Conventions of 1949. When these forces are under UN control, they are referred to as MCDA.

Other Deployed Forces. These are all military and civil defense forces deployed in the region other than MCDA. They include the forces deployed by the Affected State and any foreign forces deployed under bilateral agreements or under the auspices of organizations other than the UN.

Last Resort: Foreign MCDA should be requested only where there is no comparable civilian alternative and only the use of MCDA can meet a critical humanitarian need. The MCDA must therefore be unique in capability and availability.

3.4 International Response

International response efforts are conducted by a diverse group of organizations and entities. The independence of those organizations precludes the hierarchical command and control architecture that is traditionally used by the military. To those new to FDR operations, that lack of structure can create a perception of chaos often referred to as the *fog of relief*, see Figure 2-1.

Figure 3-1: Fog of Relief

Despite the perception that no one is in charge of international disaster response efforts, the Affected State, as a sovereign nation, has the primary responsibility for coordination, implementation, and monitoring of relief operations, including security for relief and disaster response personnel.

3.5 Affected State

The Affected State is an independent nation whose sovereignty and integrity shall be respected. Therefore, international relief operations shall be conducted only at the request or consent of the Affected State.

3.6 Transit State

A Transit State is a state through which disaster relief or recovery assistance personnel and supplies have received permission to pass to or from the Affected State.

3-4

3.7 Donor Nation

A donor nation provides financial, material, and personnel support to assist in the response and recovery from disasters.

> *NOTE:* In the case of a failed State or when the Affected State is unable to execute responsibilities for disaster response operations, while working with the Affected State, the UN will take a larger role in coordinating the response efforts. In all cases, the response will be coordinated through the UN orchestrated Cluster Approach. The UN Cluster Approach is described in Section 3.8.1.8.

3.8 International Participants

Military personnel participating in FDR operations should familiarize themselves with the organizational structure of the UN, Intergovernmental Organizations (IGO), and Non-Governmental Organizations (NGO).

An International Organization (IO) is any institution that operates in more than one country. An IO may be a Non-governmental Organization (NGO) or an Intergovernmental Organization (IGO).

International FDR participants, and some of the roles and responsibilities they fulfill, are delineated below.

3.8.1 International Organizations

International Organizations (IO) may be created by formal agreement between two or more governments. They may be established on a global, regional, or functional basis for wide-ranging or narrowly defined purposes. They are formed to protect and promote national interests shared by member states.

3.8.1.1 United Nations

The UN is an international organization, composed of member nations, chartered to facilitate cooperation in international law, international security, economic development, social progress, human rights, and achievement of world peace. The UN has the lead role in the international community to respond to natural and man-made disasters that are beyond the capacity of national authorities alone. The UN is a major provider of emergency relief and long-term assistance, a catalyst for action by governments and relief agencies, and an advocate on behalf of populations affected by disasters. A description of key UN personnel, elements, and agencies follow.

3-5

UN Emergency Resident Coordinator/Humanitarian Coordinator

The UN Resident Coordinator/ Humanitarian Coordinator (RC/HC) has a leading role in coordinating international relief with an Affected State, functioning as the chair of the Humanitarian Country Team, and assisting international UN humanitarian efforts in the Affected State.

UN Country Team

A UN Country Team consists of UN agencies, funds, and programs that primarily do development work, but are mobilized for disaster response missions. The UN Country Team should not be confused with a US Embassy Country Team (referred to throughout the handbook as a "Country Team"), which consists of the Ambassador/Chief of Mission and personnel assigned to a US Embassy.

Inter-Agency Standing Committee

Under the leadership of the UN Under-Secretary-General/Emergency Relief Coordinator (UN/ERC), the IASC develops humanitarian policies, agrees on a clear division of responsibility for the various aspects of humanitarian assistance, identifies and addresses gaps in response, and advocates for effective application of humanitarian principles. The IASC is a unique inter-agency standing committee that provides a strategic forum for coordination, policy development, and decision-making involving key UN and non-UN humanitarian assistance partners.

UN Disaster Management Team

In consultation with the Affected State and the UN Country Team, the UN RC/HC is expected to form a UN Disaster Management Team (UN DMT), which will prepare a disaster management plan. The UN RC/HC acts also as the focal point to ensure the effective dovetailing of relief assistance into rehabilitation and reconstruction programs. The purpose of a UN DMT is to ensure a prompt, effective, and concerted country-level response by the UN system. The team coordinates all disaster-related activities, technical advice, and material assistance provided by UN agencies, as well as take steps to avoid wasteful duplication or competition for resources by UN agencies. It is vital that the policies of the UN DMT mirror those approved by the Affected State.

UN Office for the Coordination of Humanitarian Affairs

The UN Office for the Coordination of Humanitarian Affairs (OCHA) is the arm of the UN Secretariat that is responsible for bringing together humanitarian response participants to ensure a coherent response to disasters. In addition, OCHA is responsible for ensuring the most efficient use of MCDA in humanitarian response operations by facilitating the relationship between the humanitarian and military components of a relief operation. UN OCHA is also responsible for operating the ReliefWeb and One Response websites, and manages the following resources:

UN Disaster Assessment and Coordination (UNDAC) Team. A five-person team assigned to support the Humanitarian Country Team with initial assessment and coordination of UN and other humanitarian efforts in response to a disaster. UNDAC teams normally arrive very early in a response.

On-site Operations Coordination Center (OSOCC). The OSOCC is established by the UN in the Affected State to coordinate incoming international disaster assistance.

Virtual On-site Operations Coordination Center (VOSOCC). The VOSOCC is a web-based information-sharing portal containing details of operational humanitarian response http://ocha.unog.ch/virtualosocc/.

UN Civil Military Coordination (UN-CMCoord) Section. The focal point for UN-CMCoord in the United Nations System is the Civil-Military Coordination Section (CMCS) of OCHA. In disaster response, CMCS may deploy UN-CMCoord Officers who assist in civil-military coordination. CMCS also provides a UN-CMCoord training program and supports military exercises.

UN Humanitarian Air Service (UNHAS). UNHAS is operated by the World Food Program to provide UN and other humanitarian organizations with safe and reliable air transport during disasters.

UN Cluster Approach. The UN Cluster Approach is the principle construct, utilized by the international humanitarian community, to facilitate a coordinated humanitarian response to an Affected State. The Cluster Approach ensures predictability and accountability in international responses to humanitarian emergencies, by clarifying the division of labor among organizations, and better defining their roles and responsibilities within the different sectors of response.

> **NOTE:** *Origins of the Cluster Approach. Ad-hoc and disjointed responses to multiple humanitarian emergencies prompted the UN Emergency Relief Coordinator (ERC) to initiate an independent Humanitarian Response Review in 2005. As a result of the review, the Cluster Approach was developed to address gaps and strengthen the effectiveness of humanitarian response through better coordination and collaboration between disaster response organizations.*

The UN Cluster Approach designates lead entities to coordinate response efforts for specific mission areas. Requirements are identified by participating organizations, in collaboration with the Affected State, and UN member organizations volunteer to fulfill them. Cluster leads have no authority to assign missions; they must meet requirements using coordination, collaboration, cooperation, and communication. The Cluster sector activities and respective leads are reflected in the following table:

#	Sector or Area of Activity	Global Cluster Lead	Symbol
1	Agriculture	UN Food and Agriculture Organization (FAO)	
2	Camp Coordination & Camp Management	UN High Commissioner for Refugees (UNHCR) International Organization for Migration (IOM)	
3	Early Recovery	UN Development Program (UNDP)	
4	Education	UN Children's Fund (UNICEF) and Save the Children UK (SC UK)	
5	Emergency Shelter	UNHCR and International Federation of the Red Cross/Red Crescent Society (IFRC)	
6	Emergency Telecommunications	UN Office for the Coordination of Humanitarian Affairs (OCHA) (process owner), World Food Program (WFP) (telecoms provider) and UNICEF	
7	Health	World Health Organization (WHO)	

8	Logistics	World Food Program	WFP
9	Nutrition	UN Children's Fund	UNICEF
10	Protection	UN High Commissioner for Refugees	UNHCR The UN Refugee Agency
11	Water, Sanitation & Hygiene	UN Children's Fund	UNICEF

Figure 3-2: UN Cluster Approach

Additional IOs include the North Atlantic Treaty Organization (NATO), the European Union (EU), the Organization for Security and Cooperation in Europe (OSCE), the Organization of American States (OAS), and the African Union (AU). NATO and OSCE are regional security organizations, while the EU, the AU, and the OAS are regional, common-interest organizations.

3.8.2 Intergovernmental Organizations

Intergovernmental Organizations (IGO) exist when two or more governments sign a multilateral treaty to form such a body and agree to finance its operations. As international entities that are created by states and that have physical plants, offices, personnel, equipment, and budgets, IGOs possess legal personality in international law. They can enter into agreements, conventions, and treaties; they can sue and be sued; they can possess property; and their staffs enjoy diplomatic status. An example of an IGO is the International Organization for Migration (IOM).

3.8.3 Non-Governmental Organizations

Non-Governmental Organizations (NGO) are independent, adaptable, and diverse primary relief providers. They fill a critical role in foreign disaster relief because many have long-term, sustaining missions and are frequently on scene before disasters occur—some operating in high-risk areas. They are likely to remain long after others have departed. Because of their capability to respond quickly and effectively to crises, they can lessen the need for military resources. For more information on NGOs, refer to the *Guide to Nongovernmental Organizations for the Military*, listed in Appendix F-References.

3-9

3.8.3.1 International Red Cross and Red Crescent Movement

On 22 June 2006, the ICRC announced that the International Red Cross and Red Crescent Movement adopted the Red Crystal as additional emblem for use by the national societies.

The International Red Cross and Red Crescent Movement consists of three separate and distinct entities, each with its own specific mission:

- *International Committee of the Red Cross* (ICRC) - Based in Geneva, ICRC is mandated to protect victims of armed conflict under International Humanitarian Law (aka, "Law of Armed Conflict"); the ICRC operates in conflict situations
- *International Federation of Red Cross and Red Crescent Societies* (IFRC) - Also based in Geneva, Switzerland, the IFRC acts as a secretariat for the national Red Cross and Red Crescent Societies, and assists in disaster management and response. The IFRC supports national societies in disaster situations
- *National Red Cross or Red Crescent Societies* - The 186 recognized national Red Cross or Red Crescent Societies are auxiliaries of their governments; national societies assist in both disasters and conflict situations

The DOD is most likely to interface with the IFRC and Red Cross/Red Crescent Societies during FDR operations.

CHAPTER 4 - COORDINATION, COLLABORATION, COOPERATION, COMMUNICATION, AND CULTURAL AWARENESS

4.1 Overview

This chapter highlights the importance of *coordination, collaboration, cooperation, communication,* and *cultural awareness* in achieving unity of effort in a FDR environment. The Joint Task Force (JTF) should become familiar with existing strategies and plans from the diverse FDR participants, including, but not limited to:

- Affected State
- US Ambassador and Country Team
- Geographic Combatant Command (GCC) Desk Officer
- US Agency for International Development (USAID), Office of US Foreign Disaster Assistance (OFDA), Disaster Assistance Response Team (DART)
- Army, Navy, and Marine Corps Civil Affairs teams operating in the area
- United Nations (UN), Intergovernmental Organizations (IGO), and Non-Governmental Organizations (NGO)

4.2 Coordination, Collaboration, and Cooperation

4.2.1 Coordination

US military FDR coordination efforts begin with USAID/OFDA prior to deployment and includes the exchange of Liaison Officers (LNO). In addition, overall response efforts may be enhanced by:

- Assigning an LNO to the Country Team
- Assigning an LNO to appropriate United Nations (UN) clusters

> **NOTE:** *LNOs are critical in the early stages of the operation when the various agencies involved have not yet established an effective coordination mechanism.*

- Maximizing coordination of Public Affairs Officer (PAO) at JTF, Department of State (DOS), GCC and the Office of the Secretary of Defense (OSD) to disseminate information consistently and accurately
- Integrating Information Operations into FDR efforts; this is a key to gaining and maintaining awareness of the operational environment, as well as shaping it (e.g., deterring violence, looting, and other illegal activity)

- Using the Affected State's communication infrastructure to minimize the use of US military equipment

Coordination with other intergovernmental organizations and nonmilitary partners requires all participants to effectively share information. The organization depicted in Figure 4-1 was used in Haiti disaster relief operations and provides a template for large-scale response operations.

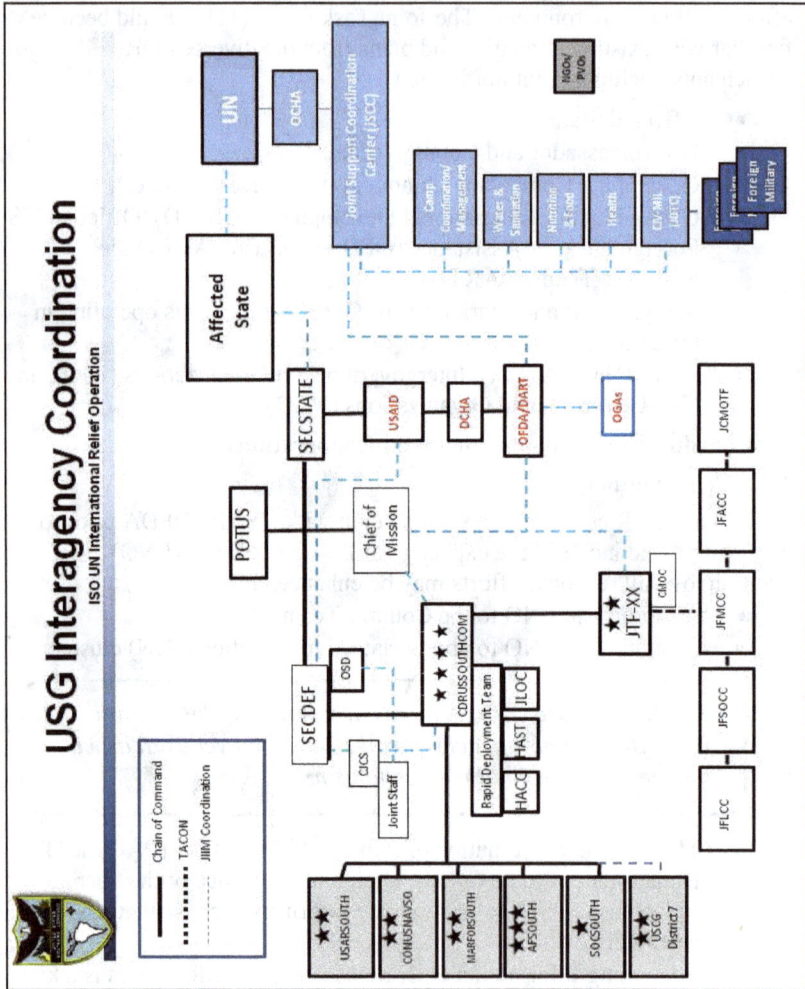

Figure 4-1: Example of USG Interagency Coordination in Support of UN International Relief Operations in Haiti

4.2.2 Collaboration

Collaboration with USAID/OFDA, the Country Team, the Affected State, UN, NGOs and IGOs, is key to developing and maintaining situational awareness, determining requirements, and supporting the response effort.

There may be response operations in Affected States where there is not a Department of Defense equivalent such as a Ministry of Defense. In such cases, the Ministry of Interior or equivalent organization may be the JTF Commander's best option for working with Affected State military forces; however, a decision in that regard should be based upon the recommendation of USAID/OFDA and the Country Team.

> "JTF Haiti operated in a complex, dynamic, permissive environment, yet an uncertain one. It included the government of Haiti, United Nations, USAID as the US Lead federal agency operating within the US Embassy and a host of interagency partners, and hundreds of NGOs. One key to JTF success was the ability to coordinate and collaborate with all the organizations. Establishing the JTF-Haiti humanitarian assistance coordination cell at the operational level facilitated this coordination and collaboration. The cell served as the conduit for bringing different organizations and functions together under one "coordination and collaboration roof."
>
> LTG P. K. Keen, USA
> Foreign Disaster Response JTF Haiti Observations
> Military Review Nov-Dec 2010

4.2.2.1 Information Sharing

Communications paths to enable the sharing of information between US and response partners must be established early to facilitate coordination prior to, and during, FDR operations.

> **NOTE:** Early in Operation Unified Response, following the earthquake in Haiti, a decision was made by the GCC commander to be open and transparent. To achieve that objective, Joint Task Force-Haiti primarily operated on unclassified systems, and used commercially available programs, such as Google Earth, to build a comprehensive common operating picture down to the tactical level.

4-3

Compliance with the DOD Unclassified Information Sharing Concept (UISC), published November 2010, of Operations (CONOPS) will substantially improve information sharing. JTFs should leverage net-centric support tools and develop architectures and processes that enhance coordination with the FDR response community.

> *The UISC CONOPS, JCS J3 dated 15 Nov 10, "... outlines the capability designed to assist joint and coalition military organizations in their efforts to collaborate, plan and coordinate operations, exchange information, and build situational awareness with both traditional and nontraditional mission partners across various mission sets.' The All Partners Access Network (APAN) was designated as the DOD UISC system of record in 2011.*

JTFs should enhance information sharing between DOD and response partners by leveraging internet-based collaborative information portals, such as the APAN and HarmonieWeb.

The APAN is an unclassified, non-dot-mil network providing interoperability and connectivity among partners over a common platform. The APAN fosters information exchange and collaboration between the United States Department of Defense (DOD) and any external country, organization, agency or individual that does not have ready access to traditional DOD systems and networks.

HARMONIEWeb is a portal site built to allow government and non-government organizations to collaborate in the areas of humanitarian assistance, disaster response, and stability and reconstruction. Users can request portal sites to meet the collaborative needs of a given situation. Once a site is created, users manage access, provide content, and designate their own administrators or site owners. See Appendix D for additional information on HARMONIEWeb.

> **NOTE:** *Accurate information sharing is governed by policies such as DODI 3000.05, DODI 8220.02, DODD 3000.07 all of which mandate the exchange of information.*

The JTF should clear information for release to non-DOD FDR participants through the Foreign Disclosure Officer (FDO). The Public Affairs Officer (PAO) should adhere to standard procedures before releasing information to the public. The PAO is responsible for obtaining Public Affairs guidance from higher authorities.

> **NOTE:** *Information intended for public release is to be vetted through the PAO and the FDO, where appropriate. Controlled Unclassified Information (CUI) and Classified Military Information (CMI) must be reviewed. Once release authority is granted from OSD through the GCC, PAOs are the release authority for all UNCLASSIFIED material intended for public release. FDOs and Designated Disclosure Authority (DDA) are the release authority for all CUI and CMI that has been properly de-classified by a DDA prior to disclosure.*

4.2.3 Cooperation

A key element of cooperation in FDR environments is understanding international partners, which begins with cultural awareness. Cultural awareness, information sharing, common tactical operating picture, internal and external communications are vital to creating and maintaining favorable public perception and positive relationships with our international partners.

The following may assist in communicating with various organizations and the JTF:

- Increase awareness and encourage attendance at JTF meetings, briefings, and joint planning sessions
- Humanitarian relief agencies may interpret responses, such as "we'll try" or "if possible" as affirmative, and subsequently expect the military to fulfill the request; if the JTF is unable to provide the support desired, relationships might be adversely affected
- Some international respondents may have charters that do not allow them to collaborate with armed forces; the JTF Commander may find a third party useful in establishing liaison
- Be aware that aid organizations may not be familiar with military structure, chain of command, or approved activities in the FDR mission
- Clearly articulate the role of the military to all participants; it is important to communicate that only limited types of support may be allowed
- Be cognizant of legal constraints

- Memorandums of agreement and understanding (these may include air/surface transportation, petroleum products, telecommunications, labor, security, facilities, contracting, engineer support, supplies, services, and medical support) should address funding considerations and delineate authority
- Take measures to minimize the perception of favoritism of a particular relief organization(s)
- Post information on the UN's ReliefWeb, which is widely used by international respondents

> **NOTE:** *In addition to USAID/OFDA, coordination with US-based NGOs can be facilitated by the American Council for Voluntary International Action (InterAction), a consortium of over 150 private agencies that operate in 180 countries. The UN Office for Coordination of Humanitarian Affairs (UNOCHA) is another valuable resource for coordination of efforts.*

4.3 Cultural Awareness

Cultural awareness is a critical ingredient in FDR missions, the success of which may be negated through cultural insensitivity. Cultural awareness involves understanding the history, customs, and social norms of the Affected State and assisting organizations.

Understanding the regional culture and how the Affected State and other participants perceive the JTF's actions and those of the United States in general is a key element of situational awareness. Participants in relief operations should be prepared to convey a non-judgmental attitude toward local customs, beliefs, and practices. Neither insensitivity nor extreme nationalism is acceptable in DOD personnel.

> **NOTE:** *Seemingly, insignificant details can have enormous cultural implications. As examples, some countries prefer white or clear pouches for human remains (body bags), and object to standard military issue olive or black pouches. Certain countries do not accept foodstuffs grown in other countries, such as yellow corn.*

JTFs should consider consulting with the following to enhance cultural understanding:

- The Country Team

- The regional and/or country desk officers of the Geographic Combatant Command (GCC)
- Foreign Area Officer (FAO)
- Army, Navy, and USMC Civil Affairs Teams (CAT)
- USAID permanent party, HAA/M, or DART
- Bi-lingual and bi-cultural advisors
- IGO/NGO personnel
- Military Information Support Operations (MISO) personnel
- Unit chaplains
- Affected State key leaders

Many cultures hold religious leaders in high esteem. Accordingly, chaplains may serve particularly well in liaison to indigenous religious leaders.

4.3.1 Language

Linguists provide a critical language and cultural awareness capability to ensure strategic messages reach the local audience. JTF Commander's should ensure assigned units have adequate linguistic support by:

- Requesting the deployment of linguists as early as possible
- Co-locating linguists with liaison officers and JTF detachments
- Seeking assistance from local first responders or other participants from the region who may be able to help with dialects, particularly in remote areas
- Drawing upon the language skills and local area familiarity of personnel attached to the JTF
- Directing subordinate commands to identify personnel with language skills and background
- As advised by USAID/OFDA and the Country Team, consider the use of pictures and other graphics as a means of communicating

JTFs should be sensitive to the use of military jargon. Some terms commonly used in military operations may have meanings or connotations that may be misinterpreted by individuals within Affected States, Inter-Governmental Organizations (IGO), and Non-Governmental Organizations (NGO).

4.4 Communications

In a FDR environment, DOD missions may require use of multiple communication systems. Communication systems will be provided by the GCC in accordance with Affected State requirements. Specific requirements may span the full capability of DOD communications abilities, including line of sight, over the horizon, and satellite communications systems for voice, video, and data. In some cases, DOD

4-7

organic communications capabilities may be the only communications infrastructure in the area of operations. These assets, whether terrestrial or satellite based, may be heavily relied upon by the Affected State and relief coordination efforts lead by the UN.

Interoperability is the key to successful disaster relief operations. Non-mission capable or incompatible communications, overloaded command centers, distraught citizens, and exaggerated or inaccurate news media coverage contribute to confusion and chaos.

Depending on the incident, a large percentage of the commercial communication system may be degraded or destroyed. Power to cell phone towers may remain problematic for an extended period. Units cannot assume that commercial wired (landline phones, etc.) or wireless communications will be functional during an incident and must plan for alternate forms of communication and power. Units should consider bringing additional satellite telephones (including dialing instructions, directory, battery charger, and case) to support civilian partner communications needs until the civilian communications grid is restored. Providing satellite telephones (with training to use them) to key civilian leaders can greatly enhance communications and coordination.

Frequencies must be coordinated through the Affected State spectrum manager. Plans for expedient communications should include:

- Liaison teams with unit compatible communications
- Satellite telephones, such as IRIDIUM, to provide voice communications
- Use of the internet
- Affected State's relationship with commercial vendors

Most communications will travel via commercial telephone networks or the internet. Signal planning must include the ability to access commercial internet, commercial telephone, and video teleconference networks. With internet access (wireless or landline) and virtual private network software, units can create a Command and Control (C2) network able to handle almost all of their requirements.

Service personnel that support civilian response efforts e.g., medical, logistics, and aviation must be able to communicate with Affected State, and international humanitarian community responders in order to coordinate relief efforts.

Units responding to support civilian responders must be prepared to integrate communication systems with civilian agencies. Because of equipment differences, spectrum requirements, and the geography at the incident, commanders should not assume that tactical radio equipment is interoperable with civilian equipment.

Interoperability planning should include radio bridging devices that can connect varied devices such as tactical radios to cell phones, and sharing data through a common information management plan.

JTFs should utilize traditional and non-traditional media to convey strategic communications messages. Traditional communications include embedding the news media to provide independent reporting to the public.

Additionally, social media can be an excellent tool for accelerating information sharing in a non-traditional manner. For example, JTF partners can use Twitter or Facebook to disseminate short situational reports and status updates.

> ⚠ *CAUTION: Social media communications can result in embarrassment or inappropriate comments being posted and attributed to the DOD. The JTF should ensure that social media is incorporated into their strategic communications plans.*

JTF Commander's may choose to establish a Joint Interagency Information Center (JIIC) to support FDR operations. Working closely with the US Embassy Public Affairs Office, the JIIC will coordinate and synchronize USG communications to ensure that they are integrated, cohesive, and timely.

4.4.1 Communications Planning

Communications planning should commence in the early stages of FDR operations. That planning must provide for an interoperable and compatible communications network that leverages organic military and commercial capabilities to support the communications requirements of the operations.

Commercial telephone networks, military satellite channels, and conventional military command and control systems will support communication of directions, orders, and information. Information protection for non-secured communications must be implemented.

Planning considerations include:
- *Communications Security* (COMSEC) is enabled through encryption or codes; COMSEC tactics may include physical security and operations security

- o Communications may be secured against monitoring whenever feasible using military or commercial encryption
- o Security may include system redundancy to reduce system failures stemming from sabotage and elements of nature
- o Communications security will also be complicated by the need to coordinate with other agencies (US and non-US) and multinational forces

- *Frequency Management* will help allocate finite frequencies, as well as accommodate multinational forces and non-military agency integration into the frequency management program
  - o Multinational forces and non-military agency integration into the frequency management program should be addressed at the GCC level
  - o Telecommunications requirements for the Affected State should also be considered
  - o The Affected State may have frequency managers who control their allocations and therefore FDR forces may not have access to some of the frequency spectrum in the Affected State (those issues should be adjudicated at the GCC level)
- *Interoperability* US operating forces may have to adapt to Affected State information systems requirements to participate effectively in FDR operations; interoperability will necessitate using unclassified communications
- *Standardize Reports* to increase efficiency of operations
- *Determining communications requirements* to provide basic and mission essential services
  - o Employ additional equipment and reconfigure connectivity to provide direct routing to principal destinations
  - o Add equipment to provide multiple routes to prevent site isolation
  - o Have sufficient equipment to support airborne capabilities, respond to new missions, and avoid critical shortages
  - o Build in redundancy to reduce system failures

4.4.2 Communicating with the International Humanitarian Community

Communications planning must consider the transfer of responsibility for communications from the United States Government to other entities such as the Affected State, UN or NGOs.

NOTE: *Most disaster relief partners, including some American and all non-American US Embassy staff, cannot access military classified material. Accordingly, in order to facilitate shared situational awareness, units should "go unclassified early."*

During the initial days following a disaster, passing timely and accurate information will be fundamental to mission success.

In the initial weeks of Operation UNIFIED RESPONSE, following the Haiti earthquake, text messages became the primary means of communication. Those messages were the simplest and most reliable means of coordinating relief efforts. The common operating picture was based upon a spreadsheet containing over 1500 data points, on everything from Internally Displaced Person camps, to medical facilities, to ration distribution sites. Utilizing that simple, near-universal format allowed the UN Office for the Coordination of Humanitarian Affairs to display the information on a Google Earth-based website, where anyone with internet connectivity could access the information.

4.5 Strategic and Crisis Communications

From the strategic to the tactical level, it is imperative that the JTF speak with one voice and act as a catalyst to facilitate unity of effort.

Joint Pub 3-13, *Joint Doctrine for Information Operations*, defines *Strategic Communication* as "focused United States Government (USG) efforts to understand and engage key audiences in order to create, strengthen, or preserve conditions favorable for the advancement of USG interests, policies, and objectives through the use of coordinated programs, plans, themes, messages and products synchronized with the actions of all elements of national power." That effort in disaster response is also known as *crisis communications*.

Figure 4-2 is a cross-comparison between DOD, DOS, and USAID/OFDA staff positions, missions, and functionality.

Agency	DOD	DOS/Embassy	USAID
Position	Public Affairs Officer (PAO)	Press Officer	Press Officer
Office	Public Affairs Offfice	Public Diplomacy Office	Legislative and Public Affairs Office
Mission	Military Information Support Operations	Public Diplomacy and Public Affairs	Legislative and Public Affairs
Function	Strategic Communications	Public Diplomacy and Public Affairs	Outreach

Figure 4-2: Staff Functionality Comparison

In its simplest form, strategic communications in disasters serve several purposes:
- Manages public expectations
- Provides public safety information
- Serves to increase the government's credibility of response
- Provides information on the relief effort to inform and calm the citizens

Neither the USG civilian agencies nor the military have control over elements of the global information environment, including:
- Mainstream broadcast
- Print journalists
- Someone on the street with a camera cell phone
- Blogger with a laptop and Internet service

The JTF must be prepared to respond in an effective and timely manner to inaccuracy, rumor, and falsehood. Minimizing inaccuracies requires integration of information activities between the Embassy, USG agencies, and the JTF.

Managing expectations of the public regarding the roles and capabilities of military forces in response to disasters is a critical part of strategic communication. That requires awareness by US military leaders of the role, methods and techniques of media; the news cycle, the competitive orientation of media outlets; the information environment; and non-traditional news sources.

4-12

In order to manage the information environment, it is essential that the Public Affairs Officer maintain situational awareness regarding the actions and locations of forces and be able to communicate their roles and missions to the media. In that way, Public Affairs Officers can keep reporters informed.

Deployment of US military forces can introduce unexpected perceptions and expectations. Inexperience with military-media relationships risks antagonism, despite the fact that crises create the shared purpose for both military and media representatives to help guide and safeguard affected populations.

> **NOTE:** Articulating the military's roles and responsibilities in disaster response to the media is critical. While opportunities can occur in formal settings, it is more important to establish informal, enduring relationships between military members (specifically the public affairs community and senior leaders) and their media counterparts.

Military leaders can improve unit effectiveness by considering non-verbal messages. For example, service-members working in individual ballistic armor with helmets and protective eyewear, even if unarmed, may send an unintended message to the populace. The same personnel in soft caps and t-shirts convey a different impression. There is no standard formula and commanders must weigh force protection requirements against perception.

As the primary communicator, the role of JTF Commander has a greater significance in a disaster relief operation. Emergencies create crises, crises demand resolution, and the engaged military leader speaking early and accurately can become a powerful symbol of positive action and crisis resolution in managing expectations and instilling confidence.

> **NOTE:** Strategic messaging helps to build confidence, capability, and goodwill in countries that have been affected by disasters. Strategic messages should be clearly articulated to the government of the Affected State and to the media. The overarching theme of those messages should be that the US response to disasters strives to serve those in need.

National-level, Affected State, and GCC strategic themes should provide the guidance from which a JTF Commander can craft strategic messages.

4-13

Examples of strategic messages include:
- The Affected State is in charge of the overall relief efforts
- The Affected State is resilient and capable of recovering from this disaster
- We are engaged in a unified effort to support the Affected State's recovery
- The USG is committed to supporting relief efforts as long as we are needed
- We are supporting Affected State first responders and the international humanitarian community
- Our focus is upon delivering food, water, and medical assistance quickly to those living in the affected areas
- We are proud to be participating in this humanitarian mission. The chance to help alleviate suffering and save lives is extremely important
- All US military efforts support the Office of US Foreign Disaster Assistance, USAID, which is the lead Federal agency coordinating US disaster relief operations
- The US military will lend its capabilities as long as they are needed
- We are one of many US government agencies working to bring the best disaster relief efforts possible to the affected population

> 66 *"Actions speak louder than words." Keep in mind that what you do, not what you say, is the only thing that will be remembered.*
> *Anonymous*

CHAPTER 5 - OPERATIONAL CONTEXT AND MISSION PLANNING CONSIDERATIONS

5.1 Overview

The Joint Operational Planning Process (JOPP) is an orderly, analytical planning process, which consists of a set of logical steps to analyze a mission; develop, analyze, and compare alternative courses of action (COAs); select the best COA; and, produce a plan or order.

This chapter is intended as a supplement to JOPP and integrates disaster response best-practices and lessons learned to facilitate planning and execution in a Foreign Disaster Relief (FDR) environment.

The nature of the operational environment will impact the conduct of FDR operations. Important elements of the operational environment considered during planning and execution phases include:

- The type of disaster involved
- The prevailing security environment

In addition to types of disasters and prevailing security environments, time (speed of onset) plays a major role in the response planning process. Regardless of the amount of time available for response planning, the information contained in this chapter has utility in both deliberate and crisis action planning.

> *Plans are worthless, but planning is everything.*
>
> *There is a very great distinction because when you are planning for an emergency you must start with this one thing: the very definition of "emergency" is that it is unexpected, therefore it is not going to happen the way you are planning.*
>
> *President Dwight D. Eisenhower*
> *14 November 1957*

5.1.1 Types of Disasters

Disasters may occur as one of two major types:

- *Slow-onset,* such as drought or famine
- *Rapid-onset,* such as tornado or earthquake

See Chapter 6 for more information on disaster typology.

5-1

5.1.2 Prevailing Security Environment

Security environments are generally comprised of two elements: the operational environment and force protection requirements.

5.1.2.1 Operational Environments

The DOD defines *operational environments* as "a composite of the conditions, circumstances, and influences that affect the employment of military forces and bear on the decisions of the unit commander." Joint Publication 1-02, DOD Dictionary provides the following definitions:

Permissive Environment – an operational environment in which the government of an Affected State has control, as well as the intent and capability, to assist response operations. That type of environment was pervasive during the relief efforts in Southeast Asia following the tsunami in 2004.

Uncertain Environment – an operational environment in which the government of an Affected State does not have effective control of the territory and population in the intended operational area. That type of environment was present during the relief efforts in Pakistan following the earthquake in 2005 and the flooding in 2010.

Hostile Environment – an operational environment in which hostile forces have control as well as the intent and capability to oppose response operations. That type of environment existed during the relief efforts in Somalia from 1992-1994.

This handbook is intended to assist operational and tactical level planners at the Joint Task Force level and below in planning the five phases of FDR operations in permissive environments.

5.1.2.2 Force Protection

Joint Pub 1-02 defines *Force Protection* (FP) as the "Preventive measures taken to mitigate hostile actions against DOD personnel, resources, facilities, and critical information. Force protection does *not* include actions to defeat the enemy or protect against accidents, weather, or disease."

Force protection is a high priority during operations. Even in a completely permissive environment, in the aftermath of a disaster the JTF can expect to encounter banditry, vandalism, and various levels of violent activities from criminals or unruly crowds. Responding forces may be particularly at risk in catastrophic-level events where they may encounter populations who are distressed and desperate for life-saving resources.

The JTF must be trained and equipped to mitigate threats to US personnel and resources. All deploying members should be provided with FP briefings prior to, and throughout, the operation.

When an operation occurs in an area that was previously torn by war or civil strife, force protection considerations may include planning for booby-traps, mines, and identification of unexploded ordnance.

> **NOTE:** *While security is the responsibility of the Affected State, Commanders must adjust FP measures based upon Affected State capabilities, the threat environment, intelligence assessments, and advice from the Country Team. If not clearly stated in the Chairman of the Joint Chiefs of Staff Execution Order, FP will be addressed in the Rules of Engagement (ROE).*

In some FDR operations, sustainment forces will require security resources to protect personnel, supplies, and equipment. Regardless of the environment, security must be factored into FP planning requirements. FP must be coordinated with the Country Team and include anti-terrorism measures appropriate for the operational environment.

5.2 Adaptive Planning

Joint Publication 5-0, *Joint Operation Planning*, defines *adaptive planning* as "the joint capability to create and revise plans rapidly and systematically, as circumstances require." Adaptive planning provides the means to ensure timely and effective responses to changes in policy and direction. Adaptive planning provides the mechanism to get the right force at the right time in the right place.

5.2.1 Mission Analysis

The primary purpose of mission analysis is to understand the problem and purpose of the operation and issue appropriate guidance to drive the rest of the planning process.

5.2.1.1 Mission-Shaping Considerations

When preparing for FDR operations, all participating organizations should consider the *strategic implications of tactical actions*. Those mission-shaping considerations are:

Global Perception. Natural disasters that affect large populations draw the immediate attention of the media. Response operations by US forces may be closely scrutinized by the international press. Global perception of the US government and its military may be shaped by this media coverage.

Medical and Public Health. The common denominator of all disasters is casualties. Accordingly, one of the most important aspects of the disaster response mission is to save lives and reduce suffering. One of the JTF Commander's priority information requirements (PIR) should include assessing (in coordination with USAID/OFDA) the medical and public health needs of the population.

Sustainable Human Development. Access to potable water, food and agricultural products, adequate shelter, medical and public health resources, education, and an adequate industrial base are all aspects of sustainable human development that may be disrupted or destroy when disasters occur.

Socio-Economic Impact. Disasters are immediately disruptive to the social and economic stability of Affected States, particularly if they are developing nations or fragile states. Response and recovery efforts at the tactical level may have an immediate and significant impact in enabling the Affected State to recover from a catastrophic event.

Geo-Political Impacts. Disasters may destabilize an Affected State's government and have negative geo-political impacts on a regional basis. Additionally, there is potential for an authoritarian regime to manipulate response efforts to assert further control over a repressed population. Care must be taken so that DOD response assets are not used in a manner contrary to US goals and ideals.

National/Trans-National Impacts. Disasters that occur in developing or fragile states can have cascading effects that may spread into surrounding countries throughout a region. This may result in trans-national implications that are difficult to manage. JTF commanders need to have an understanding of regional issues and how potential cascading effects may complicate tactical level operations.

Environmental Impacts. Disasters often have a significant and lasting effect on the environment. Environmental damage can disrupt critical sustainment resources, such as food and water. Mitigating a disaster's environmental impact may be a significant mission for the JTF.

Transition. The military provides unique capabilities to FDR operations. When military assets are requested, their use should be limited in duration and scope, and they should operate in direct support of the Affected State, the Department of State, and USAID/OFDA and in coordination with NGOs and UN agencies. Upon acceptance of a mission assignment from USAID/OFDA, the JTF Commander should ensure the JTF staff develops a transition plan or exit strategy. However, it is important to note that some missions will not transition they just end when the unique capability provided by DOD is no longer required.

5-4

Enabling Sustainable Recovery. The methodology and manner of relief operations can enable rapid recovery or, conversely, prolong the length of the recovery. Every action in an emergency response will have a direct effect on the cost of reconstruction. Sustainable recovery depends on restoring the affected population's capacity to meet its basic food, health, shelter, water, and sanitation needs. Disaster victims place a high priority on restoring their means of livelihood. Consequently, understanding the Affected State's priorities and providing assistance that supports the overall efforts to restore viable socioeconomic systems are critical.

5.2.1.2 Key Assumptions

- The Affected State has requested or is willing to accept assistance from the USG
- The Affected State has the lead and USG response will support the Affected State and tie into the UN cluster system (if established)
- USAID is the Lead Federal Agency for USG response
- DOD participation in FDR operations will be at the request of the DOS, in coordination with USAID in support of strategic objectives
- All operational and tactical level goals and objectives will be consistent with USG strategic policy objectives
- The Chairman of the Joint Chiefs of Staff (CJCS) Execution Order (EXORD) has been issued
- FDR operations will be conducted in a permissive environment; ROE will be developed in accordance with current information
- Deploying Global Response Forces will be under the operational command of the GCC
- DOD forces will be in direct support of the US Embassy/Chief of Mission (COM), and USAID/OFDA, or Disaster Assistance Response Team (DART)
- Some NGO, IGO and IOs may remain in the Affected State after the DOD departs

5.2.1.3 Conducting Assessments

Assessment is the first and most critical phase in the operational planning process. All-source estimates, area assessments, and surveys will all contribute to disaster situational awareness.

When the military arrives at the scene of a disaster, it should coordinate with the US Embassy, the USAID/OFDA DART and other non-DOD assessment teams already on-scene to obtain assessments conducted prior to their arrival to avoid duplication of effort. Those assessments focus upon political, cultural, economic, military, geographic, topographic, climatic, infrastructure, engineering, and health.

Both Geographic Combatant Commanders (GCC) and JTF Commanders may elect to establish assessment teams comprised of key staff elements. These assessment teams are usually referred to as a Humanitarian Assistance Survey Team (HAST) for more information on HAST see Section 9.2.2. JTF staff should assess the situation from a resource requirements perspective.

Furthermore, it is essential that the JTF Commander quickly identify areas where DOD may be unique qualified to conduct assessments, such as runways, seaports or waterways. Every effort should be made to reconcile differences between DOD and other USG agency assessments. Assessment information sharing may facilitate interagency coordination, cooperation, and communication, as well as, assist the GCC and JTF Commander in determining how to best utilize available military assets.

Additional information sources include: Civil Affairs Teams operating in the area, the Country Team, HAA/M at the GCCs, GCC country books, recent USAID/OFDA situation reports, and UN agency assessments. NGOs, IGOs, USG agencies, and the media are likely to have already established a presence in the Affected State. See Appendix D for useful websites. These entities, along with traditional military information sources can provide the following:

- **First-hand reports.** Forces in direct contact with affected populations are capable of providing first-hand reports that are a result of direct observation of the impact of the disaster. Those reports should provide information that can only be obtained on the scene, such as the mood of the people, their high-demand requirements, security issues, and potential cultural sensitivities.
- **Intelligence.** Consult the intelligence community for photographs, country reports, and analysis. During disaster relief operations, intelligence can play an important role in surveying the extent of damage and can assist in planning for the deployment of relief forces.
- **Media reports.** Review US and international media reports. The media is a key source of information in developing and maintaining situational awareness.
- **Commercial imagery.** Commercial imagery (e.g., Google Earth) is generally of sufficient quality and granularity for use and carries few, if any, restrictions on distribution. Note that commercial imagery databases may not reflect current conditions or infrastructure.

- **ReliefWeb and the Virtual On-Site Operations Coordination Centre (Virtual OSOCC).** These online resources facilitate continuous and simultaneous information exchange between governments and organizations responding to disasters. An account is required.
- **Photographs.** Combat Camera can be an important resource in capturing key still and video images to aid in the assessment process and provide historical visual references of the disaster and its associated impact.
- **Lessons learned.** Search the DOD Joint Lessons Learned Information System (JLLIS) database (and other USG agencies)for relevant information. Lessons learned based on previous operations, exercises, and country visits can help develop a better understanding of FDR mission requirements.

Initial DOD assessments should provide baseline data that can be used as a reference for further monitoring and evaluation. Assessment is a continual process that occurs throughout the course of an operation and needs to be updated based on emerging operational conditions. Subsequent assessments should build on previous findings and expand the information database.

5.2.2 Functional Area Assessments

FDR operations normally require assessments from a variety of specialties including personnel from engineering, health service support (HSS), civil affairs (CA), and logistics to help with operational planning and force positioning. The following paragraphs discuss some considerations for specific functional area assessments.

5.2.2.1 Health Service Support Assessments

Essential functions of HSS during the assessment phase include: conducting a heath situation estimate, determining mission requirements, and considering how to enhance force health protection and operational stress control (OSC). Knowledge of health risk factors in the environment US forces will be operating in will assist in mitigating health risk to US forces. Understanding the capabilities of contributing organizations, including those of the Affected State, helps to reduce duplication and ensures efficiency of HSS efforts. Together this information serves to ensure medical resources are structured efficiently to meet the needs of the deployed force and to support the JTF's tasking and commander's intent.

5.2.2.2 Logistics Assessments

Lessons learned indicate that the logistics support necessary to sustain an FDR operation are frequently underestimated. Logistics planners assess DOD-organic and Affected State capabilities. Logistics objectives and

5-7

mission risks should also be identified and assessed. Emphasis must be placed upon locating logistic bases as close as possible to the relief recipients. All potential supply sources should be considered, including Affected State, commercial, multinational, NGOs, and pre-positioned supplies.

5.2.2.3 Engineering Assessments

Engineering personnel can assist with infrastructure inspection, damage assessments, and the identification of repair requirements to Affected State facilities and critical infrastructures. Engineering assessments are vital to ensuring response operations are not impeded.

5.2.2.4 Risk Assessment

Risk assessments identify the specific hazards that the joint force may encounter during the mission, and determine the probability and severity of loss linked to those hazards. See Chapter 8 for details on this process.

5.2.3 Assessment Categories

Assessment information normally falls into one of two major categories:

- Situational assessment (environment)
- Needs assessment (people)

The *situation assessment* collects information about the effects of the disaster. This assessment gathers information on the magnitude of the disaster and the extent of its impact on the population and the physical infrastructure, as well as the environment.

The *needs assessment* indentifies resources and services for immediate emergency measures to save and sustain the lives and property of the affected population.

NOTE: *Do not delay life-saving activities while awaiting comprehensive assessments.*

Gathering information for situational and needs assessments can be done concurrently. Information collected during the initial assessments will form the basis for determining the type and amount of relief needed during the immediate response phase of the disaster.

Areas assessed include:

- Priority and required scale of response
- Type, duration, method, and location(s) of assistance
- Degree and nature of local support to the response

As a result of ongoing assessments, the mission requirements may undergo major changes during the operation.

Distinguishing between emergency and chronic needs is the first step; assessments must differentiate between what is normal for the location and what is occurring as a result of the disaster. Disaster response efforts must address the most dire needs of the affected population. Care must be taken not to delve into mission areas that would otherwise qualify as sustainable human development initiatives and nation building operations.

5.2.3.1 Example Situation Assessment (Environment)

The following are examples of functional areas that require assessments in FDR operations:

- Institutions
  - Affected State
    - Leadership status
    - Police-Security
    - Military
  - International Organizations
  - DOS (US Embassy)
  - Communications capability
  - NGO's present and their mission/capabilities
  - Military Assistance Advisory Groups
  - Social Cohesion
  - Confidence in Affected State Government
    - View of political leadership
    - View of police
    - View of military
  - Unity of Affected State social organizations
    - Religious institutions
    - Civic groups
  - Social solidarity
    - Factionalism
    - Racism
    - Sexism
    - Religious tension
    - Cultural inhomogeneity
  - View of foreigners

- Communications
    - Affected State/local
    - International
    - Security
    - JTF Headquarters/Components
        - Tactical/secure
        - Commercial
- Transportation (Road, Rail)
    - Affected State vehicle availability
        - Emergency vehicles
        - Buses
    - Equipment requirements
    - Other sources
    - Road, bridges, railroads
    - Transportation security
- Airfields
    - Locations & capacity (primary/alternates)
        - Ability to conduct Visual Flight Rules/ Instrument Flight Rules (VFR/IFR)
        - What's required
    - Flight Standards - International Civil Aviation Organization (ICAO)/Federal Aviation Administration (FAA)
    - Fueling (type and availability)
        - Jet A/Jet –A1
        - Unleaded
        - Diesel
        - Motor gasoline (MOGAS)
    - Services (type and availability)
        - Support equipment
        - Communications

The Central Intelligence Agency *World Factbook* is an excellent source of information on the location, size, and capacity of worldwide airports and seaports and should be used as a primary information resource during assessment operations.

- Ports
    - Locations & capacity (primary/alternates)
    - Ability to conduct port operations
    - Services
        - Support equipment

- Communications
  - Port/cargo handling personnel
  - Storage/warehousing capacity/capability
    - Security
  - Fueling (type and availability)
    - Jet A/Jet –A1
    - Unleaded
    - Diesel
    - Motor gasoline (MOGAS)
- Miscellaneous
  - Weather conditions (typical seasonal and long-range forecasts)
  - Topography and hydrography
  - Dangerous flora and fauna (including marine threats)
  - Local pollution and prevalence of hazardous material (including nuclear/radiological)

5.2.3.2 Example Needs Assessment (People)
- Medical
  - Type of major injuries/endemic or recent disease threats
    - Affected State medical capabilities/degradation
    - External expertise specialties required (numbers/types)
    - Other sources (NGOs/IGOs)
  - Location and type of existing health facilities
  - Emergency medical facilities
  - Medical equipment (available/required)
  - Medical supplies (available/required)
  - Affected State blood requirements
- Water
  - Production/purification capabilities and capacity
    - Municipal
    - Other water treatment systems
  - Requirements
    - Available potable sources (wells/surface/subsurface)
  - Quality/portability
  - Distribution capacity
- Food
  - Availability/requirements
    - Type and quantities required

5-11

- - Special cultural and/or religious sensitivities related to foodstuffs
  - o Storage/warehousing capacity
  - o Distribution capability/requirements
- Shelter
  - o Affected State
    - Accommodations available/required
    - Tent city sites (if required)
    - Locations (sports fields/arenas, parks, etc.)
  - o Accommodations for FDR personnel
    - Availability
    - Transportation
    - Communications
    - Security
    - Working order of hotel services (e.g., latrines, showers, heating or AC, water)
- Language and Culture
  - o Translator and transcription requirements
  - o Cultural and religious sensitivities
  - o Special practices
  - o Special observances
  - o Protection of Holy Sites
  - o Protection of cultural and historical antiquities
- Military
  - o How can the military's assets and capabilities best be utilized
  - o What military assets are in need
  - o What military assets are redundant
  - o Are there limitations to the use of military assets (e.g., no landing zones for helicopters)

5.2.4 Mission Statement Development

JTF commanders consider several factors in developing their mission statement, including: the GCC's mission statement; forces available; the operational environment; and, security considerations. As an example, the SOUTHCOM EXORD 16 January 2010 (Haiti) stated: "Execute FDR ops in coordination with and in support of USAID/OFDA and State in order to mitigate near term human suffering and accelerate recovery."

5.2.5 Concept of Operations Development

Concepts of Operations (CONOPS) concisely express what the Commander intends to accomplish and why. It describes how actions of components and supporting organizations will be integrated, synchronized, and phased to accomplish the mission, including potential branches and sequels.

For the purposes of this handbook, the CONOP consists of five operational phases:

Phase 1 – Preparation for Deployment
- Begins with JTF activation and receipt of deployment order
- Ends with deployment of assigned forces

Phase 2 – Deployment
- Begins with deployment of forces to vicinity of the foreign disaster
- Initial deployment ends when sufficient forces are in place to begin effective relief operations

Phase 3 – Foreign Disaster Relief Operations
- Begins with commencement of FDR activities
- Ends when mission objectives are met or transition ordered

Phase 4 – Transition
- Begins with execution of a transition plan
- Ends when DOS, DOD, or Affected State declares that US forces are no longer required

> **NOTE:** *Sometimes missions end without a transition phase because the Affected State no longer requires, or is no longer requesting, DOD support.*

Phase 5 – Redeployment
- Begins as forces start to redeploy
- Ends when all deployed assets return to home station or other designated area

5.2.6 Operations Order

Once a course of action has been selected, JTF Commander will release an Operations Order (OPORD), which directs supporting commands and organizations to execute the mission. As forces begin to deploy and perform assigned tasks, operational metrics will be used to measure effectiveness of the mission.

5.2.7 Operational Metrics

Metrics matter. Metrics are the means by which operational progress is measured. Metrics capture and demonstrate level of effort/need and measures of performance/effectiveness. Relevant metrics facilitate accurate and timely reporting to higher echelon commands and national authorities. It is important that the metrics utilized by the JTF be consistent with those used by the US Embassy and USAID/OFDA. Data collection requirements and the associated standardized metrics should be disseminated to deploying forces as early as possible.

The JTF Commander and Staff should not develop their own metrics, but instead use internationally accepted metrics. The SPHERE Project developed a handbook entitled the *"Humanitarian Charter and Minimum Standards in Disaster Response"*. The SPHERE initiative was launched in 1997 by a group of NGOs and the Red Cross and Red Crescent movements, and identified minimum standards to be attained in disaster assistance in each of 6 key sectors—water supply and sanitation, nutrition, food aid, food security, shelter, and health services. To date, over 400 organizations in 80 countries have contributed to the development of minimum standards and key indicators, making it the most comprehensively universal resource for the application of disaster assistance metrics. An additional resource for defining minimum standards is the USAID Field Operations Guide (FOG) for Disaster Assessment and Response. Minimum standards are located on the inside cover. The SPHERE Handbook and the FOG are both available on-line, respectively, at:

http://www.sphereproject.org/component/option,com_docman/task,cat_vie w/gid,70/Itemid,203 (available in multiple languages); and,

http://www.usaid.gov/our_work/humanitarian_assistance/disaster_assistanc e/resources/pdf/fog_v4.pdf

5.2.7.1 Types of Metrics

When assessing FDR operations, four different categories of metrics may be collected and evaluated: level of need (LON), level of effort (LOE), measure of performance (MOP), and measure of effectiveness (MOE). Examples of each category are provided below.

Level of Need. The needs of the population should be estimated across a variety of categories, depending on the nature of the disaster and the situation on the ground. Determining LON metrics could demand subject matter expertise not resident in the military, and requiring forces to use estimates developed by USAID/OFDA DART, United Nations (UN), NGOs, or other agencies.

Level of Effort. LOE measures quantify the magnitude of specific efforts and are the easiest for units to collect and report. These measures are more helpful to analysts if they include specific details, such as what was delivered, by what means, and to what destinations. Metrics should use standard units, such as weight (e.g., kilograms or pounds), volume (e.g., liters of water), or contain all the appropriate variables to calculate real units (e.g., "2 pallets; 4000 bottles per pallet; 500 ml water per bottle," to allow analysts to calculate a total of 400 liters of water).

Measures of Performance. MOP provide insight into how efficiently tasks are being performed. These measures are usually related to mission essential task accomplishment at the unit or group level. MOP often can stand alone as measures of how efficiently the force is performing specific assigned tasks, though they may not provide insight into progress the force is making toward overall mission accomplishment.

Measures of Effectiveness. MOE are calculated based upon indications of how well mission objectives are being met. In order to calculate a MOE, one will need estimates of both the LOE (how much has been done to help) as well as the LON (how much help is needed). If the LOE metric and LON metric are describing the same variable in the same units, the MOE is usually calculated as a percentage or fraction (LOE/LON) of the need that is being met. If tactical forces are not able to provide accurate needs assessments with their own resources, it can be helpful to develop MOE with the intelligence directorate and to employ disaster assessment and surveillance teams to establish levels of need.

In some cases, unit commanders may be unable to report MOE due to uncertainty about the needs in their area of operations. However, all units can report LOE and MOP. Tactical-level reporting of LOE and MOP remains valuable because it can be used by higher headquarters to calculate operational MOE based on high-level assessment of need reported by civilian agencies.

Metrics by Type

Level of Need   Civilian Needs
- Daily food and water needed in an area
- Number waiting for transportation
- Number in need of shelter or clothing
- Number with a particular medical condition

Military Needs

- Number of pallets of material in need of transportation (e.g., from the forward logistics site to the APODs)
- Number of Meals Ready to Eat (MRE) required daily to feed military personnel in-country
- Number of gallons of gasoline needed daily in the JOA to sustain military relief operations on the ground

Level of Effort

Resources/Capabilities Expended
- Number of personnel in a camp
- Daily sorties
- Daily flight hours
- Daily fuel expended
- Passengers transported
- Liters of water delivered
- Number of medical conditions treated

Measures of Performance

Efficiency of Task Accomplishment
- Time for augmentation personnel traveling into theater to get to their assigned station in the JOA
- Rate at which a port or harbor is cleared
- Percent of cargo shipped into the Logistics Operations Area (LOA) that is being tracked to its recipient (e.g., in-transit visibility)
- Time for refrigerated blood shipments to be transported from the originator to medical facilities in the JOA

Measures of Effectiveness - Usually calculated as a percentage or fraction (LOE/LON) of the need that is being met

5.2.8 Mission End State

For specific situations that require the employment of military capabilities, the President, Secretary of State, and the Secretary of Defense typically will establish strategic objectives. Achievement of those objectives should result in attainment of the desired end state. Based upon strategic guidance, the JTF Commander will tailor CONOPS to meet the desired end state.

CHAPTER 6 - DISASTER TYPOLOGY

6.1 Overview

The United Nations (UN) utilizes the following definition of a *natural disaster*:

> " *A natural disaster is the result of a vast ecological breakdown in the relationship between man and his environment; a serious and sudden (or slow, as in drought) disruption on such a scale that the stricken community needs extraordinary efforts to cope with it, often with outside help or international aid.*

6.2 Types of Disasters

Disasters are grouped into three classes: *natural* (e.g. seismic, meteorological); *man-made* (e.g., a result of technology, explosions, or fires); and *mixed* (a combination of natural and man-made). Additionally, disasters are classified by the speed of onset, i.e., rapid or slow (see Section 5.2.1).

6.2.1 Earthquakes

Earthquakes are caused by the discharge of energy accumulated along a geological fault. This discharge of energy varies in magnitude. The underground point of origin of the earthquake is called the "epicenter." Earthquakes by themselves rarely kill people. However, the secondary events they trigger—such as building collapses, fires, tsunamis (seismic sea waves), and volcanoes—actually cause the majority of injuries and death.

Earthquakes often occur with little or no warning. Frequently there will be *aftershocks*, smaller earthquakes that happen for days and sometimes weeks after the major earthquake. Aftershocks are a concern for units providing support in response to a disaster.

Another dangerous effect of earthquakes is liquefaction of the soil. *Liquefaction* occurs when the structure of loose, saturated soil breaks down due to some rapidly applied loading. As the structure breaks down, the

Credit: US Navy

Figure 6-1: The Collapsed Presidential Palace of Haiti

6-1

loosely-packed individual soil particles attempt to move into a denser configuration. In an earthquake, however, there is not enough time for the water in the pores of the soil to be squeezed out. Instead, the water is "trapped," and prevents the soil particles from moving closer together. This is accompanied by an increase in water pressure that reduces the contact forces between the individual soil particles, thereby softening and weakening the soil deposit.

In an extreme case, the pore-water pressure may become so high that many of the soil particles lose contact with each other. In such cases, the soil will have very little strength, and will behave more like a liquid than a solid— hence, the name "liquefaction".

6.2.1.1 Measuring Earthquakes

Earthquakes have traditionally been measured by the "*Richter Scale*," developed in 1935 by Charles Richter. The Richter Scale doesn't accurately measure the size of earthquakes higher than 8.0, so the *Moment Magnitude Scale* was introduced in 1979 as a replacement. This is now the official scale used by the United States Geological Survey (USGS) and seismologists around the world.

> **NOTE:** *The media and public information officers associated with official agencies still say "Richter Scale," but are actually reporting measurements from the Moment Magnitude Scale.*

Earthquake magnitude is a logarithmic measure of earthquake size. In simple terms, this means that at the same distance from the earthquake, the shaking will be 10 times as large during a magnitude 5.0 earthquake as during a magnitude 4.0 earthquake. The total amount of energy released by the earthquake, however, goes up by a factor of 32.

The smallest earthquakes that are likely to be noticed are quakes at 3.0 on the scale. Earthquakes of 4.0 rattle objects, but normally do not cause appreciable physical damage. Earthquakes of 5.0 might cause some amount of building damage, but rarely cause loss of life. Earthquakes of 6.0 in intensity can cause major damage and substantial loss of life. Depending on soil density and other geological features, an earthquake measuring 7.0 or higher on the Richter scale will result in massive destruction and high levels of casualties and death, if it occurs in a populated area.

6.2.1.2 Impact and Related Hazards

The biggest hazards after an earthquake are:

- Fire as a result of broken gas lines and sparks from frayed electrical lines
- Flooding as a result of nearby reservoirs or damns rupturing or disruptions to underground water mains
- Electrical shock as a result of downed power lines
- Structural collapse as a result of fragile infrastructure or substandard construction
- Power and communications systems failure
- Exposure to toxic industrial chemical/materials released due to failure of infrastructure or ruptured storage containers
- Diarrheal and communicable disease due to poor sanitation and contaminated water sources

WARNING: *Always assume downed power lines are live. Water (including snow) is an excellent conductor of electricity. Stay away from downed power lines of any kind to avoid electrical shock and grave injury.*

Be aware of your surroundings and observe what you are walking under or near at all times.

NOTE: *After an earthquake, people may be trapped in collapsed structures and debris. Rapid deployment of trained search and rescue teams with dogs and high-tech listening devices are necessary to locate and save them.*

The most common injuries in an earthquake are:
- *Crush injuries*
- *Traumatic injuries*
- *Burns*
- *Smoke inhalation*

6.2.1.3 Earthquake Missions

Regardless of preparation by the Affected State, a major earthquake may quickly overwhelm the ability to respond. A JTF Commander tasked to support Affected State authorities may receive the following resource requests:

- Aviation (fixed and rotary wing) assets to conduct search and rescue, personnel transport/recovery, transport logistics, or perform aerial structural damage assessment

- Engineer support to clear roads or create emergency bypasses, clear rubble, and clear ports and waterways of hazards to navigation
- Key infrastructure assessments (US Army Corps of Engineers or Naval Facilities Engineer Command qualified personnel)
- Large-deck naval platforms for mass movement of evacuees
- Logistical support such as water, food, temporary shelters, and generators
- Material handling equipment and trained operators
- Medical evacuation assets (air and ground)
- Medical support personnel, deployable medical treatment facilities, Class VIII (medical) materiel
- Shoring materiel and debris clearing equipment
- Transportation

6.2.2 Landslides

The term *landslide* includes all types of gravity-induced mass movements of earthen material, ranging from rock falls through slides/slumps, avalanches, and flows. It also includes both sub-aerial and submarine mass movement triggered mainly by precipitation (including snowmelt), seismic activity, and volcanic eruptions.

For simplification, the term *debris flow* will include mudflows, debris torrents, and lahars (volcanic debris flows). Debris flows are fast-moving landslides that can occur in a wide range of environments. A debris flow is a rapidly moving mass of water and material that is mainly composed of sand, gravel, and cobbles, but typically includes trees, cars, small buildings, and other loosened debris material. A debris flow normally has the consistency of wet concrete and can move at speeds in excess of 35 mph.

Landslides occur when the stability of a slope changes from a stable to an unstable condition. A change in the stability of a slope can be caused by a number of factors, acting together or alone. Natural causes of landslides include:

Figure 6-2: Landslide in El Salvador

- Groundwater pressure acting to destabilize the slope

- Loss or absence of vegetation and soil structure (e.g. after a wildfire)
- Erosion of the base of a slope by rivers or ocean waves
- Weakening of a slope through saturation by snowmelt, glaciers melting, or heavy rains
- Earthquakes adding loads to unstable slopes
- Earthquake-caused liquefaction destabilizing slopes
- Volcanic eruptions

Landslides can be aggravated by human activities, such as: deforestation, cultivation, and construction, which may destabilize already fragile slopes. Other aggravating activities include:

- Vibrations from machinery or traffic
- Blasting
- Earthwork which alters the shape of a slope, or which imposes new loads on an existing slope

6.2.2.1 Impact and Related Hazards

- Fire as a result of broken gas lines
- Flooding from disruptions to underground water mains
- Electrical shock as a result of downed power lines
- Structural collapse
- Secondary landslides and localized earth settling
- Release of hazardous material
- Disease outbreaks can occur when spores are spread in dust clouds
- Power and communications systems failure

6.2.2.2 Landslide Missions

Regardless of preparation by an Affected State, a major landslide may quickly overwhelm response capabilities. Several major landslides in the 1980s were responsible for eliminating entire villages in South America. A JTF Commander tasked to support Affected State authorities may receive the following resource requests:

- Aviation (fixed and rotary wing) assets to conduct search and rescue, personnel transport/recovery, transport logistics, or perform aerial structural damage assessment
- Engineer support to clear roads or create emergency bypasses, clear rubble, and clear ports and waterways of hazards to navigation
- Key infrastructure assessments (US Army Corps of Engineers or Naval Facilities Engineer Command qualified personnel)

- Logistical support such as water, food, temporary shelters, and generators
- Material handling equipment and trained operators
- Medical support personnel, deployable medical treatment facilities, Class VIII (medical) materiel
- Medical evacuation assets (air and ground)
- Shoring materiel and debris clearing equipment
- Transportation

6.2.3 Tsunami

A Tsunami is a series of waves formed by displaced water due to the disturbance of the sea floor that enters coastal areas, often causing serious damage. *Tsunami* (pronounced tsoo-nah-mee) is a Japanese word that literally means, "harbor wave." Tsunamis are appropriately named so because they are silent and unseen in the ocean waters, but are often fierce in shallow coastal waters. They are often mistakenly referred to as *tidal waves*. Tidal waves are caused by the gravitational pull of the moon and sun, and tend to be much higher vertically then a tsunami. Tsunamis are usually the result of underwater seismic activity, and are sometimes referred to as *seismic sea waves*. Tsunamis can also be caused by other non-seismic events and usually occur during the following two major geologic movements:

Figure 6-3: Tsunami Waves Hitting Japan

- Underwater landslides
- Underwater volcanic explosion

There are three types of tsunamis: local, regional, or distant.

- *Local tsunamis* are generally located very close to a landmass; these types of tsunami can be very dangerous due to the availability of very little warning

6-6

- *Regional tsunamis* are generated between one and three hours travel time away from their destination; they are generally the most common, but tend to cause the least damage due to oceanographic conditions that limit the magnitude of the tsunami
- *Distant tsunamis* are generated from a long way away, such as from across the Pacific; they are generally less common, but have the potential for causing catastrophic damage

Tsunami wave speed is dependent on ocean depth: local tsunami waves travel slower (~60 mph) than deep ocean waves (up to 500 mph). Despite their speed, deep ocean tsunamis often go unnoticed because their waves may only reach a couple of feet on the sea surface. When a tsunami comes closer to a coast, wave heights will typically increase as the ocean becomes shallower.

NOTE: *People who have survived tsunamis often describe them as long "walls of water." On shore, what is noticed first depends on what part of the waves reaches land first (i.e., the wave crest causes a rise while a wave trough causes a recession of water). The terminology used to measure the height of the water above the sea level is called "run-up." Tsunamis may reach a maximum run-up on shore of 100-150 feet.*

6.2.3.1 Impact and Related Hazards

- Tsunami wave trains (generated when an oceanographic disturbance causes a horizontal wave through the ocean, similar to ripples formed by tossing a rock into a pond, spreading outward from the disturbance in 5 to 90 minute intervals)
- Drowning
- Strong currents and debris
- Flooding
- Electrical shock and fires as a result of damaged utilities
- Structural collapse
- Landslides and mudslides/localized earth settling
- Release of hazardous material
- Diarrheal and communicable disease due to poor sanitation and contaminated water sources
- Increased risk of wound infections and subsequent gangrene secondary to exposure to fecal and other contaminants in raised water tables
- Power and communications systems failure

> ⚠️ **WARNING:**
> - *Local tsunamis can follow a seismic event very quickly, leaving less time for evacuation and incident management preparation*
> - *Early returns of people to their settlements without anticipating the wave train or to witness the tsunami can result in significant numbers of injuries and deaths*
> - *The issuance of a tsunami warning means that evidence exists for a potentially destructive tsunami and evacuation is strongly advised*
> - *On the other hand, a watch status means that a tsunami may have been generated, the wave travel time is more than 3 hours, and evacuation is needed if the watch is upgraded to a warning*
> - *For tsunamis of distant origin, the danger areas are 50 feet above sea level and within 1 mile of the coast*
> - *For tsunamis of local origin, potential danger areas are those less than 100 feet above sea level and within one mile of the coast*

6.2.3.2 Tsunami Missions

Regardless of the degree of civil preparedness, a major tsunami may quickly overwhelm the response capability of the Affected State. A JTF Commander may receive the following resource requests:

- Aviation (fixed and rotary wing) assets to conduct search and rescue, personnel transport/recovery, transport logistics, or perform aerial structural damage assessment
- Engineer support to clear roads or create emergency bypasses, clear rubble, and clear ports and waterways of hazards to navigation
- Key infrastructure assessments (US Army Corps of Engineers or Naval Facilities Engineer Command qualified personnel)
- Large-deck naval platforms for mass movement of evacuees
- Logistical support such as water, food, temporary shelters, and generators
- Material handling equipment and trained operators

- Medical support personnel, deployable medical treatment facilities, Class VIII (medical) materiel
- Medical evacuation assets (air and ground)
- Shoring materiel and debris-clearing equipment
- Transportation

6.2.4 Hurricanes, Cyclones, and Typhoons

Figure 6-4: Hurricanes Jeanne, Karl, and Tropical Storm Lisa

The terms *hurricane* and *typhoon* are regionally specific names for a strong *tropical cyclone*. A *tropical cyclone* is the generic term for a non-frontal low-pressure system over tropical or sub-tropical waters with organized convection (i.e., thunderstorms) and cyclonic surface wind circulation.

- *Tropical depressions* have sustained winds of less than 39 mph
- *Tropical storms* have sustained winds from 39-73 mph
- *Cyclones* have sustained winds above 74 mph and are referred to as:
  - *Hurricane* (in the North Atlantic Ocean, the Northeast Pacific Ocean east of the dateline, or the South Pacific Ocean east of 160E)
  - *Typhoon* (in the Northwest Pacific Ocean west of the dateline)
  - *Severe Tropical Cyclone* (in the Southwest Pacific Ocean west of 160E or Southeast Indian Ocean east of 90E)
  - *Severe Cyclonic Storm* (in the North Indian Ocean)
  - *Tropical Cyclone* (in the Southwest Indian Ocean)

6-9

Tropical cyclones are one of nature's most powerful forces. A tropical cyclone's winds blow in a spiral direction around a relatively calm area known as the *Eye*. The eye is usually 20 to 30 miles wide. The most violent activity takes place in the area immediately around the eye, called the *Eyewall*.

As these storms approach from the ocean, the skies begin to darken and the wind gets stronger. As it nears land, it may bring torrential rain, storm surges, and very high winds. A single tropical cyclone can last for more than 2 weeks in open waters. The heavy rain brought by these storms not only threatens coastal areas, but can also hit hundreds of miles inland, in some cases causing flooding that occurs days after the storm actually hits shore.

6.2.4.1 Measuring Tropical Cyclones

The Saffir-Simpson Scale, Figure 6-5, is the American methodology for classifying the intensity of tropical cyclones. Other classification scales exist in geographically specific locations throughout the world.

Category	Winds	Effects
One	74-95 mph	No real damage to building structures. Damage primarily to unanchored mobile homes, shrubbery, and trees. Some coastal road flooding and minor pier damage.
Two	96-110 mph	Some roofing material, door, and window damage to buildings. Considerable damage to vegetation, mobile homes, and piers. Coastal and low-lying escape routes flood 2-4 hours before arrival of center. Small craft in unprotected anchorages break moorings.
Three	111-130 mph	Some structural damage to small residences and utility buildings with a minor amount of curtain wall failures. Mobile homes are destroyed. Flooding near the coast destroys smaller structures with larger structures damaged by floating debris. Terrain continuously lower than 5 feet Above Sea Level (ASL) may be flooded inland 8 miles or more.
Four	131-155 mph	More extensive curtain wall failures with some complete roof structure failure on small residences. Major erosion of beach. Major damage to lower floors of structures near the shore. Terrains continuously lower than 10 feet ASL may be flooded requiring massive evacuation of residential areas inland as far as 6 miles.

Five	Greater than 155 mph	Complete roof failure on many residences and industrial buildings. Some complete building failures with small utility buildings blown over or away. Major damage to lower floors of all structures located less than 15 feet ASL and within 500 yards of the shoreline. Massive evacuation of residential areas on low ground within 5 to 10 miles of the shoreline may be required.

Figure 6-5: Saffir-Simpson Hurricane Wind Scale

6.2.4.2 Impact and Related Hazards

- Tornadoes and waterspouts
- Strong currents and debris associated with storm surge
- Flooding
- Electrical shock and fires as a result of damaged utilities
- Structural collapse
- Landslides and mudslides/localized earth settling
- Erosion of roads, beaches, and coastlines
- Release of hazardous material
- Power and communications systems failure
- Drowning
- Contaminated water supplies and exposure to waterborne disease (cholera, shigella, salmonella, and the hepatitis A virus)
- Increased risk of wound infections and subsequent gangrene secondary to exposure to fecal and other contaminants in raised water tables

> *NOTE: Critical infrastructure such as hospitals, electrical generation facilities, sewage treatment plants, food production facilities, etc., are just as prone to being damaged and/or rendered unusable as any other building or facility in the impacted area.*

6.2.4.3 Hurricane, Cyclone, and Typhoon Missions

Regardless of preparation by the Affected State, a major tropical cyclone may quickly overwhelm the ability to respond. A JTF Commander tasked to support Affected State authorities may receive the following resource requests:

- Assets to clear ports and waterways of hazards to navigation

- Aviation (fixed and rotary wing) assets to conduct search and rescue, personnel transport/recovery, transport logistics, or perform aerial structural damage assessment
- Engineer support to clear roads or create emergency bypasses, clear rubble, and clear ports and waterways of hazards to navigation
- Key infrastructure assessments (US Army Corps of Engineers or Naval Facilities Engineer Command qualified personnel)
- Logistical support such as water, food, temporary shelters, and generators
- Material handling equipment and trained operators
- Medical support personnel, deployable medical treatment facilities, Class VIII (medical) materiel
- Medical evacuation assets (air and ground)
- Shoring materiel and debris clearing equipment
- Transportation

NOTE: *Mass fatality management resources may be a significant requirement in cyclonic events that are rated either extreme or catastrophic (Categories 4 & 5).*

6.2.5 Tornadoes

Figure 6-6: Double Tornado

A tornado is a violently rotating column of air extending from swelling clouds to the ground. The most violent tornadoes are capable of tremendous destruction with wind speeds of 250 mph or more. Damage paths from a tornado can be in excess of 1 mile wide and 50 miles long.

Tornadoes strike with little or no warning and can occur with equal frequency in all countries. Tornadoes may appear nearly transparent until dust and debris are picked up or a cloud forms within the funnel. Tornadoes have been known to move in any direction. The average forward speed is 30 mph but may vary from nearly stationary to upwards of 70 mph. *Waterspouts* are tornadoes that form over warm water. They can move onshore and cause damage to coastal areas.

6-12

6.2.5.1 Measuring Tornadoes

Tornado severity is measured by the Enhanced Fujita Tornado Scale (see Figure 6-7). This scale ranges from designations of EF0 to EF5. Originally designed as a measurement of structural damage caused by a tornado, the scale is now most often used to describe a tornado's wind speed.

Category	Winds	Effects
EF0	65-85 mph	Small trees uprooted, broken tree limbs
EF1	86-110 mph	Roofs damaged, mobile homes uprooted
EF2	111-135 mph	Roofs removed, mobile homes demolished
EF3	136-165 mph	Roofs and walls torn down, all trees uprooted
EF4	166-200 mph	Homes leveled, foundations moved
EF5	Over 200 mph	Buildings destroyed, cars thrown > 100 meters

Figure 6-6: Enhanced Fujita Tornado Scale

6.2.5.2 Impact and Related Hazards

- Debris
- Electrical shock and fires as a result of damaged utilities
- Structural collapse
- Release of hazardous material
- Loss of geographic reference or local landmarks
- Localized power and communications systems failure
- Contaminated water supplies and exposure to waterborne disease (cholera, shigella, salmonella, and the hepatitis A virus)
- Increased risk of wound infections and subsequent gangrene secondary to exposure to fecal and other contaminants in raised water tables

> ⚠️ **WARNING:** *Tornado winds can propel objects in excess of 200 mph. A 2x4 piece of wood moving at 150 mph is capable of penetrating an 8-inch concrete block wall. Small objects, such as nails, can cause massive penetrating injuries at tornado speeds (nail guns, in comparison, only shoot nails at 70 mph).*
>
> *Tornados may be accompanied by lightning, hail, and rain, all capable of causing dangerous, life-threatening conditions on their own.*

6.2.5.3 Tornado Missions

Although tornadoes occur worldwide, these destructive forces of nature are found most frequently in the United States. Unless a rare "super-cell" occurs causing multiple tornadoes to touch down nearly simultaneously, it is unlikely that the DOD would be asked to respond to an event outside of the United States. In the isolated case(s) that a request for support is made, the mission requirements would be similar to those associated with tropical cyclones.

6.2.6 Wildfires

According to the United States Geological Survey (USGS), a wildfire is any uncontrolled fire in combustible vegetation that occurs in the countryside or a wilderness area. Other names such as brush fire, bushfire, forest fire, grass fire, hill fire, peat fire, vegetation fire, veldfire, and wildland fire may

Credit: NOAA

be used to describe the same phenomenon depending on the type of vegetation being burned.

A wildfire differs from other fires by its extensive size, the speed at which it can spread out from its original source, its potential to change direction unexpectedly, and its ability to jump gaps such as roads, rivers and fire breaks. Wildfires are characterized in terms of the cause of ignition, their physical properties such as speed of propagation, the combustible material present, and the effect of weather on the fire.

Figure 6-7: Wildfire

Wildfires occur on every continent except Antarctica. While some wildfires burn in remote forested regions, they can cause extensive destruction of homes and other property located in the *wildland-urban interface* (i.e., the zone of transition between developed areas and undeveloped wilderness).

The four major natural causes of wildfire ignitions are lightning, volcanic eruption, sparks from rock falls, and spontaneous combustion. However, many wildfires are attributed to human sources such as arson, discarded cigarettes, sparks from equipment, and power line arcs.

6.2.6.1 Impact and Related Hazards

- Smoke inhalation
- Falling embers
- Explosions as a result of damaged utilities
- Structural collapse
- Release of hazardous material
- Localized power and communications systems failure
- Loss of geographic reference or local landmarks

WARNING: *Even healthy people can have respiratory problems from the particulates and gasses in the air when wildfires occur. More intense or prolonged exposure can also cause chest pain, shortness of breath, and headaches. Those with pre-existing condition such as asthma, bronchitis, or chronic obstructive pulmonary disease (COPD) can have their conditions severely exacerbated to the point of requiring acute medical care and/or life-support.*

The real danger from smoke is toxic gasses. Carbon monoxide is the most common, but there can be complex and more dangerous gasses created by toxic compounds burning downwind.

6-15

> ⚠️ **CAUTION:** *Wildfires are very destructive to property, but the bigger danger to humans and animals is poor quality air and smoke that can reach populations well beyond the geographic scope of the wildfire.*
>
> *Wildfires have a rapid forward rate of spread when burning through dense, uninterrupted fuels. They can move as fast as 6 mph in forests and 14 mph in grasslands.*
>
> *Re-ignition poses a serious risk to response assets conducting wildfire incident management. Careful consideration should be given to evacuation plans to facilitate rapid egress from re-ignited wildfires.*
>
> *Wildfires cannot be controlled with a garden hose. Personnel involved with incident management efforts should leave wildfire management to professionally trained and equipped fire fighters.*

6.2.6.2 Wildfire Missions

Regardless of preparation by the Affected State, a major wildfire can quickly overwhelm the ability to suppress the fire and respond to support an affected population. A JTF Commander tasked to support Affected State authorities may receive the following resource requests:

- Air assets for search and rescue (SAR), personnel transport/recovery, medical evacuation (MEDEVAC), logistics transport, or aerial structural damage assessment
- Aviation (fixed and rotary wing) assets to conduct fire suppression
- Debris clearing equipment
- Key infrastructure assessments (US Army Corps of Engineers or Naval Facilities Engineer Command qualified personnel)
- Logistical support such as water, food, medical supplies, temporary shelters, and generators
- Military personnel and engineering equipment
- Transportation of first responders, evacuees, displaced personnel, injured, medically fragile or special needs populations

6.2.7 Floods

Floods are the most common natural disasters, accounting for approximately 40% of all disasters worldwide. More than 90% of the world's population lives within ten miles of a major body of water and is

6-16

thus subject to the effects of flooding. Areas of greatest risk of flooding include low-lying areas; coastal communities; and those areas located downstream from dams. Floods may be caused by an abundance of rainfall; melting snow; or the expanding development of wetlands, which reduces absorption of rainfall.

Credit: USAID

Figure 6-8: Flood in Bangladesh

Flash floods can occur within six hours of a rain event, after a dam or levy fails, or the sudden release of water from an ice or debris jam. Flash floods are the number one cause of natural disaster-related deaths. Floods are often long-term events and may last days, weeks, or longer.

> ⚠ **CAUTION:** *Failing to promptly evacuate at-risk populations from a flood watch area will exacerbate response requirements. It is easier to re-locate relatively stable and healthy populations than it is to accommodate their life support requirements under conditions of severe environmental duress.*

6.2.7.1 Related Hazards

- Drowning
- Hypothermia
- Strong currents and floating debris
- Electrical shock and fires as a result of damaged utilities
- Structural collapse
- Landslides and mudslides/localized earth settling
- Release of hazardous material

6-17

- Power and communications systems failure
- Loss of geographic reference or local landmarks
- Animal and reptile bites, and increased risk of vector-borne illness
- Contaminated water supplies and exposure to waterborne disease (cholera, shigella, salmonella, and the hepatitis A virus)
- Increased risk of wound infections and subsequent gangrene secondary to exposure to fecal and other contaminants in raised water tables
- Unsanitary living conditions

> **WARNING:** *The appearance of shallow waters in flooded areas can be extremely deceiving. Response vehicles entering flood waters erroneously presumed to be safe can be swept away with the force of the flow in as little as two feet of water. As little as four inches of water can sweep a human off their feet and result in drowning.*
>
> *Floods can continue to be dangerous, even after the water has receded, due to various forms of contamination (e.g., fecal bacteria from sewers, corrupted hazardous waste, and various forms of chemical and fuel contaminants). Extreme care must be exercised by response personnel in flood environments.*

6.2.7.2 Flood Missions

A JTF Commander tasked to support Affected State authorities may receive the following resource requests:

- Air and water-borne assets for search and rescue (SAR), personnel transport/recovery, medical evacuation (MEDEVAC), logistics transport, or aerial structural damage assessments
- Assisting in set-up of temporary staging areas (indoor and outdoor) and temporary storage areas
- Assisting in constructing temporary shelter for disaster responders; displaced, affected civilians; and emergency services personnel
- Assisting in constructing temporary sites in proximity to the disaster site for medical support or evacuation transfer, communications node set-up/operation, electrical power generation, and logistical support operations
- Deployable medical treatment facilities, equipment, and Class VIII (medical) supplies

- Medical providers and public health support personnel
- Engineering support may include:
  - Assessing damage to roads, bridges, structures, utilities, and other critical infrastructure (USACE/NAVFAC)
  - Clearing debris and mud
  - Conducting topographic surveys for the extent of flood damage
  - Constructing temporary bridges
  - Opening roadways for emergency and medical traffic
  - Overprinting maps to depict damage, water levels, key facilities, and SAR search and rescue activities
  - Providing emergency power and/or restoring power to critical facilities
  - Restoring critical facilities, services, and utilities
- Providing expedient repair of critical distribution systems
- Supporting evacuation of seriously ill or injured patients to locations where hospital care or outpatient services are available
- Supporting points of distribution for food, water, and medical supplies
- Transportation assets

6.2.8 Volcanic Eruptions

The most common type of volcanic eruption occurs when magma (the term for lava when it is below the Earth's surface) is released from a volcanic vent. Eruptions can be *effusive*, where lava flows like a thick, sticky liquid; or *explosive*, where fragmented lava explodes out of a vent. In explosive eruptions, the fragmented rock may be accompanied by ash and gases; in effusive eruptions, degassing is common but ash is usually not.

Figure 6-9: Hawaiian Eruption, Fire Fountaining at the Pu'uO'o Volcano in Hawaii

An *explosive eruption* blasts solid and molten rock fragments (tephra) and volcanic gases into the air with tremendous force. The largest rock fragments (bombs) usually fall back to the ground within 2 miles of the vent. Small fragments (less than about 0.1 inch across) of volcanic glass, minerals, and rock (ash) rise high into the air, forming a huge, billowing eruption column. Eruption columns can grow rapidly and reach more than 12 miles above a volcano in less than 30 minutes, forming an eruption cloud.

Volcanic gases and ash clouds in the troposphere and lower stratosphere can pose a serious hazard to aviation. Large eruption clouds can extend hundreds of miles downwind, resulting in ash fall over an enormous area. Heavy ash fall can collapse buildings, and even minor ash fall can damage crops, electronics, and machinery.

Fumaroles (small openings co-located with cracks in the ground) allow gases to reach the surface, even when a volcano is not erupting. More than 90% of all gas emitted by volcanoes is water vapor (steam), most of which is heated ground water. Other common volcanic gases are carbon dioxide, sulfur dioxide, hydrogen sulfide, hydrogen, and fluorine. Sulfur dioxide gas can react with water droplets in the atmosphere to create *acid rain*, which causes corrosion and harms vegetation. Carbon dioxide is heavier than air and can be trapped in low areas in concentrations that are deadly to people and animals. Fluorine, which in high concentrations is toxic, can be adsorbed onto volcanic ash particles that later fall to the ground. The fluorine on the particles can poison livestock grazing on ash-coated grass and contaminate domestic water supplies.

> **NOTE:** Cataclysmic eruptions, such as the June 15, 1991 eruption of Mount Pinatubo (Philippines), inject huge amounts of sulfur dioxide gas into the stratosphere, where it combines with water to form an aerosol (mist) of sulfuric acid. By reflecting solar radiation, such aerosols can lower the Earth's average surface temperature for extended periods of time by several degrees Fahrenheit (°F). These sulfuric acid aerosols also contribute to the destruction of the ozone layer by altering chlorine and nitrogen content in the atmosphere.

6.2.8.1 Types of Eruptions

Volcanologists classify eruptions into several different types (e.g., Hawaiian, Strombolian, and Plinian). Some are named for particular volcanoes, others the resulting shape of the eruptive products or place where the eruptions occur. Figure 6-10 shows a Hawaiian eruption. Figure 6-11 show different types of eruption features used to classify a volcano.

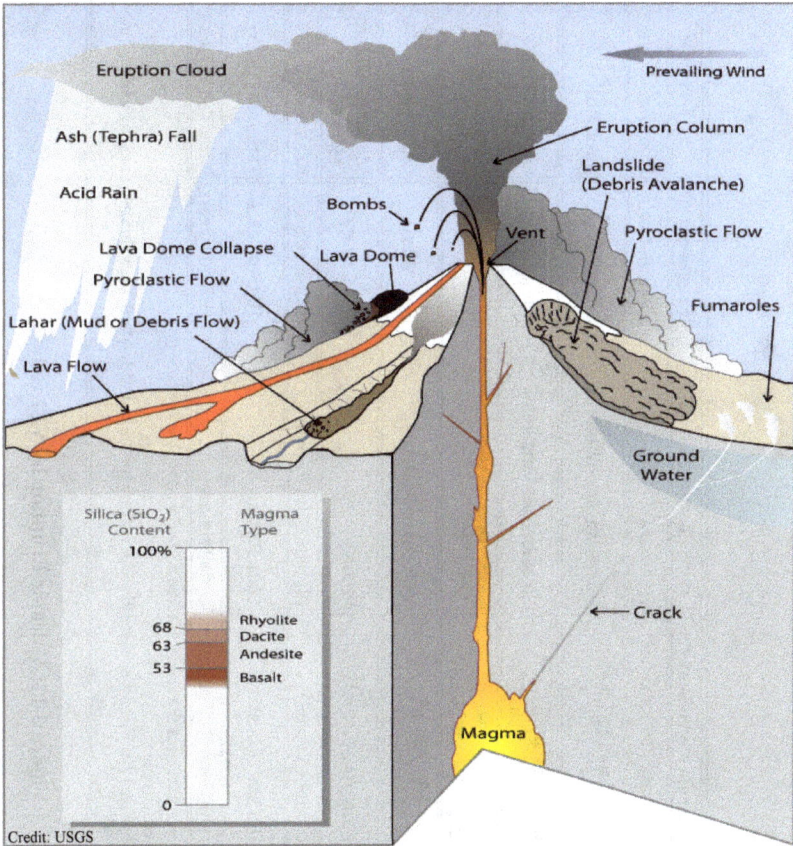

Figure 6-10: Eruption Features

6.2.8.2 Measuring Volcanic Eruptions

The Volcanic Explosivity Index (VEI), shown in Figure 6-12, is a scale for measuring the strength of eruptions used by the Smithsonian Institution's Global Volcanism Program to assess the impact of historic and prehistoric lava flows. The VEI operates similar to the Richter scale for earthquakes, in that each interval in value represents a tenfold increase in magnitude.

VEI	Plume height	Eruptive volume *	Eruption type	Frequency **	Example
0	<100 m (330 ft)	1,000 m³ (35,300 cu ft)	Hawaiian	Continuous	Kilauea
1	100–1,000 m (300–3,300 ft)	10,000 m³ (353,000 cu ft)	Hawaiian/Strombolian	Months	Stromboli
2	1–5 km (1–3 mi)	1,000,000 m³ (35,300,000 cu ft) †	Strombolian/Vulcanian	Months	Galeras (1992)
3	3–15 km (2–9 mi)	10,000,000 m³ (353,000,000 cu ft)	Vulcanian	Yearly	Nevado del Ruiz (1985)
4	10–25 km (6–16 mi)	100,000,000 m³ (3.53×10^9 cu ft)	Vulcanian/Peléan	Few years	Galunggung (1982)
5	>25 km (16 mi)	1 km³ (0.24 cu mi)	Plinian	5–10 years	Mount St. Helens (1980)
6	>25 km (16 mi)	10 km³ (2 cu mi)	Plinian/Ultra Plinian	1,000 years	Krakatoa (1883)
7	>25 km (16 mi)	100 km³ (20 cu mi)	Ultra Plinian	10,000 years	Tambora (1815)
8	>25 km (16 mi)	1,000 km³ (200 cu mi)	Ultra Plinian	100,000 years	Lake Toba (74 ka)

* This is the minimum eruptive volume necessary for the eruption to be considered within the category.
** Values are a rough estimate. Exceptions occur.
† There is a discontinuity between the 2nd and 3rd VEI level; instead of increasing by a magnitude of 10, the value increases by a magnitude of 100 (from 10,000 to 1,000,000).

Figure 6-12: Volcanic Eruption by VEI Index

6.2.8.3 Impacts and Related Hazards

- Fire
- Lava
- Toxic gasses (sulfur dioxide, hydrochloric acid, etc.)
- Asphyxiation
- Explosive eruption and falling debris
- Structural collapse
- Landslides and mudslides/localized earth settling
- Flash flooding due to rapidly melted snow and ice
- Release of hazardous material
- Damage to fixed and rotary wing aircraft operating in airspace
- Contamination of food and water sources
- Power and communications systems failure
- Loss of geographic reference or local landmarks
- Lightning and electrocution
- Tsunamis and earthquakes

WARNING: *In the absence of visible volcanic activity, gas released from lake water turnover could be a sign of an imminent eruption and on its own may produce injuries from superheated steam or bleed-off of noxious gases.*

Dormant or suspected dormant volcanoes can erupt with little or no warning.

Because of the tremendous forces associated with volcanoes, even communities thought to be out of harm's way may be suddenly imperiled.

The greatest single measure of improving survivability in volcanic eruptions is distancing. To the greatest extent possible, seismic early warning systems (EWS) that provide indications of volcanic activity should be leveraged to distance populations-at-risk from an event.

6.2.8.4 Missions Involving Volcanic Eruption

Volcanic eruptions can occur with little or no warning, regardless of the preparation by the Affected State; a major eruption is therefore likely to overwhelm the ability to respond to support an affected population. A JTF

Commander tasked to support Affected State authorities may receive the following resource requests:

- Air assets for search and rescue, personnel transport and recovery, medical evacuation, logistics transport, or aerial structural damage assessment
- Debris clearing equipment
- Deployable medical treatment facilities
- Resources for temporary staging bases
- Heavy equipment and trained operators
- Identification of intermediate staging bases for evacuated populations-at-risk
- Engineer support to clear roads or create emergency bypasses, clear rubble, and clear ports and waterways of hazards to navigation
- Key infrastructure assessments (US Army Corps of Engineers or Naval Facilities Engineer Command qualified personnel)
- Large-deck naval platforms for mass movement of evacuees
- Logistical support such as water, food, medical supplies, temporary shelters, and generators
- Medical providers and public health support personnel
- Transportation of first responders, evacuees, displaced personnel, injured, medically fragile, or special needs populations

CHAPTER 7 - PLANNING AND EXECUTION

7.1 Overview

The purpose of this chapter is to provide the Joint Task Force (JTF) Commander and staff with a list of suggested activities and actions that will enable them to successfully plan and execute a FDR operation. Information contained in this handbook has been developed based on a compilation of best practices and lessons learned from across the Geographic Combatant Commands (GCC) and the Services.

7.2 Joint Task Force Organization

The JTF is the most common type of organizational structure used by the Department of Defense (DOD) for FDR. The authority establishing the JTF determines the command relationships and assigns missions and forces. A JTF is routinely assigned a Joint Operations Area (JOA) in the GCC's Area of Responsibility (AOR). JTFs normally operate at the operational level; however, they maintain responsibility for actions that occur at the tactical level.

The unique aspects of interagency, Intergovernmental Organization (IGO), and Non-governmental Organizations (NGO) coordination and collaboration require the JTF Commander to be especially flexible and responsive. Depending on the type of contingency operation, the extent of military operations and degree of interagency involvement, coordination with civilian agencies may occur at the embassy, the JTF HQ, the Civil-Military Operations Center (CMOC), or UN Cluster meetings.

> **NOTE:** *Members of the JTF must have a clear understanding of the nature and amount of support they will be allowed to provide. Affected States, international respondents, and NGOs may view the military as an inexhaustible resource, and may inundate the CJTF with requests for support. Normally, requests for support will come to DOD through DOS at the Executive Secretary level. If circumstances on the ground dictate that the JTF has authority to fill certain types of requests from those organizations, the granting of that authority, and guidance on its use, will be included in the execute order.*

7.2.1 JTF Commander

The JTF Commander's duties may include the following:

- Making recommendations on the proper employment of assigned and attached forces
- Supervising all aspects of the JTF's planning efforts including preparation of operation orders (OPORD), and time-phased force and deployment data (TPFDD)
- Establishing requisite policies and guidelines; to include the establishment of offices, cells, elements, centers, boards, working groups, and planning teams
- Applying risk management across the full range of military operations
- Exercising or delegating appropriate, operational control (OPCON) over assigned and attached forces; the Commander may also exercise tactical control (TACON), administrative control (ADCON), have coordinating authority, or be in a supported or supporting relationship

- Providing guidance to subordinate and supporting forces
- Keeping higher HQs informed on progress of ongoing operations and mission status
- Ensuring staff conducts coordination with other forces and agencies assigned or attached, including Affected State ministries, NGOs, and IGOs
- Exercising directive authority for support capabilities delegated by the higher HQs
- Ensuring that cross-service support agreements are in place and that forces operate as mutually supporting teams
- Identifying requirements for additional forces or personnel to higher HQs

7.2.1.1 Commander Key Tasks

❑ Provide commander's guidance on establishing liaison with participating agencies (at national and local levels)
❑ Establishing liaison with Country Team/Military Group
❑ Establishing dialogue with USAID/OFDA
   o GCC Humanitarian Assistance Advisor/Military (HAA/M)
   o Disaster Assistance Response Team (DART)
❑ Conduct risk management processes
❑ Determine advance party requirements
❑ Ensure Staff Judge Advocate (SJA) is fully engaged in all aspects of mission planning
❑ Have SJA provide staff briefings on existing Status of Force Agreement (SOFA), international treaties, specific issues of international law regarding the Affected State, and FDR legal authorities
❑ Determine external relationships
❑ Establish communication protocols, use military assets for internal communications, and develop specific plans for civilian communications
❑ Ensure method is established to track mission costs
❑ Plan for media interactions; develop strategic messages for the mission and ensure promulgation to all levels
❑ Provide commander's guidance on standards of conduct and expectations of behavior
❑ Determine Liaison Officer (LNO) requirements
❑ Complete mission assurance requirements (steps to safeguard personnel and equipment)
❑ Achieve and maintain 100% personnel accountability

- ❏ Ensure compliance with force protection requirements and established Force Protection Condition (FPCON) levels
- ❏ Determine information requirements: Commander's Critical Information Requirements (CCIR), Priority Information Requirements (PIR), Friendly Force Information Requirements (FFIR), and Essential Elements of Friendly Information (EEFI)
- ❏ Ensure staff
  - o Develops a plan to address house, feed and protect unit personnel
  - o Determines source of water, power, shower, and laundry
  - o Sends small deployable communications packages ahead of larger deployable command posts for immediate feedback of requirements
  - o Considers how accidents and incidents will be processed
- ❏ Brief personnel on the humanitarian principles underlying FDR operations
- ❏ Develop strategic message for the mission and ensure promulgation to all levels
- ❏ Plan for media and VIPs
- ❏ Post deployment
  - o Conduct equipment inventories, inspections, and initiate adjustment documents
  - o Complete all investigations, particularly those concerning injuries (Line of Duty investigations), vehicle accidents, and lost or damaged equipment
  - o Consolidate costs associated with requests for assistance submitted by USAID/OFDA/DART using the Mission Tasking Matrix "MITAM" form
  - o Prepare draft After Action Report (AAR) and lessons learned

7.2.2 Deputy Commander/Chief of Staff

The Deputy Commander/Chief of Staff (DC/COS) is the key staff integrator. The DC/COS coordinates staff actions, processes, and procedures. DC/COS duties may include (but are not limited to) the following:

- ❏ Serving as the principal assistant to the Commander
- ❏ May serve concurrently as chief of staff
- ❏ Performing duties as directed by the Commander (CDR) including:
  - o Representing the CDR

      o   Assuming command if the commander becomes incapacitated or unavailable to exercise command
      o   Chairing designated boards
      o   Supervising staff planning
      o   Supervising designated subordinate unit activities or functions

- ❑ Representing the CDR when authorized
- ❑ Implementing policies as directed by the CDR
- ❑ Coordinating and directing the staff actions
- ❑ Supervising preparation of staff estimates, plans, and orders
- ❑ Establishing and monitoring the battle rhythm to ensure it effectively supports planning, decision-making, and other critical functions. See Figure 7-1 for a Notional Battle Rhythm.

Battle Rhythm Event Title	LEAD	JTF supporting members															
		COS	J1	J2	J33	J35	J4	J6	J9	JAG	PAO	Surgeon	Assessments	FP	STRATCOM	POLAD	HAA/M
Intel Analyst Sync	Intel BW		X		X												
Joint Operations Center Turnover Brief	BWC	X		X	X	X	X	X		X	X						X
Commander's Update Brief	J33	X	X	X	X	X	X	X	X	X	X	X	X				X	X
Air Planning Board	Air Lead			X	X	X	X										X
COCOM Mission Update Brief	J3	X		X	X	X	X			X	X						X	X
Joint Collection Working Group	J2			X	X	X						X					
Strategic Communications Board	J35	X	X	X	X	X	X	X		X	X						X
Joint Logistics Working Group	J4	X			X		X		X								X
Intel Analyst Sync	Intel BW	X		X													
Rules of Engagement Working Group	DJ3	X			X	X				X			X		X	X	
Joint Planning Group	J35	X	X	X	X		X	X	X				X				X
Mitam Working Group	J3	X		X	X	X	X	X	X	X		X			X	X	X
Joint Operations Center Turnover Brief	BWC	X															X
Commander's Decision Brief	COS	X	X	X	X	X	X	X	X	X	X	X	X				X	X
Joint Transportation Board	J4	X		X	X	X	X			X			X		X		X	X
Joint Acquisition Review Board	J4	X					X			X							
Effects Assessment Working Group	Dep Asmt	X		X		X	X	X	X	X	X	X			X	X	X
Joint Interagency Working Group	J9	X			X	X	X		X	X	X	X	X		X	X	X
Force Flow Working Group	J4		X		X	X	X						X			X	

Figure 7-1: Conceptual Battle Rhythm

- ❑ Formulating and disseminating staff policies
- ❑ Ensuring liaison is established with the higher, adjacent, and subordinate HQ, and other agencies and organizations
- ❑ Supervising staff, managing facilities and resources
- ❑ Supervising staff training and integration programs

7.2.2.1 DC/COS Preparation for Deployment

- ❑ Manage staff functions

- ❏ Lead staff through Joint Operational Planning Process
  - o Conduct Mission Analysis
  - o Develop Courses of Action (COA)
  - o Analyze COA
  - o Compare COA
  - o Select/Modify COA
  - o Prepare the Commander's Estimate
- ❏ Ensure procedures are in place to capture costs
- ❏ Ensure each staff element initiates a chronological staff journal for all actions in support of the crisis
- ❏ Verify recall roster information with J-1 "battle roster"
- ❏ Establish a dialogue with Embassy/Defense Attaché (DATT)
- ❏ Determine requirements for Joint Operations Center and Tactical Operations Center (JOC/TOC) manning augmentation
- ❏ Verify establishment of information flow between JOC/TOC and Embassy/DATT
- ❏ Develop a Hazardous Materials (HAZMAT) strategy
- ❏ Begin to monitor work/rest cycles

7.2.2.2 DC/COS Deployment

- ❏ Ensure procedures are in place to capture costs
- ❏ Maintain active dialogue with Embassy/DATT
- ❏ Request copy of ExecSec Memo
- ❏ Maintain flow of information between JOC/TOC and Embassy/DATT
- ❏ Initiate actions to safeguard resources, personnel, and equipment
- ❏ Prepare command brief for VIPs
- ❏ Ensure personnel accountability and security
- ❏ Be prepared to conduct split-based operations
- ❏ Execute HAZMAT strategy and be responsible for the environment
- ❏ Monitor work/rest cycles (commanders and staff are your principal concerns)
- ❏ Ensure security posture is maintained
- ❏ Develop transition plan
- ❏ Direct staff to compile lessons learned throughout all phases of the operation

7.2.2.3 DC/COS FDR Operations

- ❏ Ensure procedures are in place to capture costs
- ❏ Maintain personnel accountability and security
- ❏ Be prepared to conduct split-based operations

❏ Continue to execute HAZMAT strategy
❏ Monitor work/rest cycles (commanders and staff are your principal concerns)
❏ Ensure security posture is maintained
❏ Continue to safeguard resources, personnel, and equipment
❏ Develop closeout and redeployment plan in coordination with Affected State, USAID/OFDA, NGOs, and IGOs as appropriate

7.2.2.4 DC/COS Transition

❏ Ensure personnel accountability and security
❏ Be prepared to conduct split-based operations
❏ Continue to execute HAZMAT strategy
❏ Monitor work/rest cycles
❏ Ensure security posture is maintained
❏ Ensure re-deployment orders, tickets, travel/transportation of personnel and equipment back to home station is coordinated
❏ Ensure staff transitions appropriate responsibilities to a counterpart
❏ Determine stay-behind personnel
❏ Finalize closeout and redeployment plan in coordination with Affected State, USAID/OFDA, NGOs, and IGOs as appropriate

7.2.2.5 DC/COS Redeployment

❏ Consolidate cost data
❏ Track re-deployment/personnel accountability
❏ Continue to execute HAZMAT strategy
❏ Ensure staff transitioned appropriate responsibilities to a counterpart
❏ Execute closeout and redeployment plan in coordination with Affected State, USAID/OFDA, NGOs, and IGOs as appropriate
❏ Prepare After Action Report (AAR) comments and document lessons learned
❏ Attend USAID/OFDA AAR

7.2.3 Command Senior Enlisted Leader
7.2.3.1 Command Senior Enlisted Leader Overview

The Command Senior Enlisted Leader (CSEL) is a key advisor to the commander on all matters related to personnel and operational issues. The CSEL can also serve as an observer of activities within the operational area for the CDR.

7.2.3.2 Command Senior Enlisted Leader Key Tasks Include:

❏ Participate in staff planning processes
❏ Performing special duties as directed by the CDR including:
    o Monitoring or observing critical subordinate unit actions

o Monitoring the discipline, morale, and mission readiness of JTF elements

❑ Establish relations with each Services' senior enlisted leaders
❑ Identify quality of life issues (food, housing, uniforms, etc.)
❑ Communicate strategic messages and themes
❑ Assist in communicating General Orders; maintain discipline and enforce standards
❑ Review Service policies and procedures
❑ Ensure personnel are recognized for outstanding service and accomplishments
❑ Serve as a sounding board for JTF Commander and staff

7.3 Joint Staff Directorates

JP 3-0, *Joint Operations* describes the basic J-code directorates of a Joint Staff. Those primary staff directorates provide staff supervision of related processes, activities, and capabilities associated with the basic joint functions. See Figure 1-4 in Section 1.5.2.3.

7.4 Joint Task Force Primary Staff Directorates

JP 3-33, *The Joint Task Force* describes the typical Joint Task Force (JTF) staff elements and functions. JTF Commanders have the latitude to add, change, combine, or reorganize J-code functions to meet specific mission requirements and environments. For example, JTF Commanders will often combine the J-3, J-5, and J-7 functions. A typical JTF staff is depicted in Figure 7-3.

Joint Directorates	Function
J-1	Manpower and Personnel
J-2	Intelligence
J-3/5/7	Operations, Plans, and Training
J-4/8	Logistics, Resources, Expeditionary Contracting
J-6	Command, Control, Communications, and Computer Systems/Information
J-9	Interagency/ Civil-Military Coordination

Figure 7-2: Typical JTF Staff Functions

Special staff (not included in this matrix) may include: Staff Judge Advocate, Surgeon, Chaplain, Comptroller, Political-Military Advisor, Public Affairs Officer, Science Advisor, and Knowledge Management.

7-8

Special Staff are usually denoted as "double digit" J-Codes (e.g. J-03/Public Affairs Officer).

7.4.1.1 Staff Action Checklists

The following staff action checklists articulate best practices for five-phased FDR operations (see Section 5.3.7 for more details on FDR phases). These tasks are applicable to operational and tactical levels.

7.4.2 J-1 Manpower and Personnel/Human Resources

J-1 is responsible for all human resources, personnel actions, tracking deployment status, casualties, and losses by unit. The J-1 should also coordinate with his or her J-1 counterpart at the GCC for replacement or augmentation personnel. Personnel management in a FDR environment presents some unique challenges including:

- Working with civilian medical facilities in tracking military personnel status
- Integrating and accounting for DOD civilians/contractors augmenting deployed forces

7.4.2.1 J-1 Preparation for Deployment

- ❑ Track and report unit personnel strength
- ❑ Determine personnel augmentation requirements (including those from external agencies) for JOC and TOC
- ❑ Conduct pre-deployment readiness checks (such as legal, medical, family, and passport)
- ❑ Establish personnel accountability procedures; track authorized, assigned, present for duty, leave, sick call, and casualties by component in accordance with higher HQ reporting procedures
- ❑ Be prepared to monitor DOS tracking of American Citizens and designated foreign nationals
- ❑ Modify data collection procedures for casualty reporting and tracking to address unique communications challenges of operating in a FDR environment
- ❑ Prepare appointment orders
- ❑ Develop procedures to manage pay issues in an FDR environment
- ❑ Manage personnel readiness programs
- ❑ Develop "leave under emergency conditions" procedures
- ❑ Plan and publish personnel tracking requirements, reports, and timelines
- ❑ Be prepared to establish mail operations at deployment plus 60 days (D+60)
- ❑ In coordination with J-4, establish Administrative and Logistics Operations Center

- ❑ Develop the Joint Reception, Staging, and Onward Integration (JRSOI) plan for deploying forces
- ❑ Develop the Joint Manning Document (JMD)
- ❑ Request award's policy and authorities for JTF Commander and below
- ❑ In coordination with the SJA, determine how mission-specific General Orders, policies, and procedures will apply to deploying civilian personnel

7.4.2.2 J-1 Deployment

- ❑ Be prepared to provide a J-1 representative for the advance party
- ❑ Be prepared to conduct JRSOI throughout mission
- ❑ Participate in staff planning and OPORD development
- ❑ Utilize Delivery Performance Achievement and Recognition System (DPARS) and Cost Assessment and Program Evaluation (CAPE)
- ❑ Ensure all military personnel update their information in the appropriate service personnel accountability and assessment systems
- ❑ Establish reporting procedures for subordinate units and coordinate with higher HQ for submission of Joint Personnel Status and Casualty Reporting (JPERSTAT)
- ❑ Develop joint manning document billets and request individual augmentee sourcing
- ❑ Maintain accountability of deployed DOD personnel in affected area

> **NOTE:** *Tracking of American citizens (AMCITS) is a DOS and US Embassy responsibility via DOS Form 88.*

7.4.2.3 J-1 FDR Operations

- ❑ Maintain accountability of deployed DOD personnel
- ❑ Participate in staff planning and OPORD development
- ❑ Submit JPERSTAT report in accordance with higher HQ guidance
- ❑ Identify POC at Embassy responsible for tracking of American Citizens
- ❑ Track US DOD casualties
- ❑ If family members of unit personnel live in the area affected by the disaster, provide information on support services
- ❑ Establish procedures for family members to contact military personnel

❑ In coordination with the US Embassy, be prepared to provide personnel to support visitor operations

❑ Process personnel awards

7.4.2.4 J-1 Transition

❑ Track deployed personnel

❑ Track units as they depart the Area of Responsibility (AOR)

❑ Track DOD casualties

❑ Manage out-processing of personnel, augmentees, and contractors

7.4.2.5 J-1 Redeployment

❑ Monitor and track redeployment of all assigned US DOD personnel

❑ Coordinate with JTF Surgeon to ensure that all assigned military personnel complete a post-deployment health assessment (DOD Form 2796) prior to leaving the AOR or upon return to home station within 30 days of redeployment

❑ Ensure completion of Post-Deployment Health Reassessments (DOD Form 2900) by all deployed individuals 90 to 180 days after redeployment to home station

❑ Process and manage personal decorations, unit citations, and appropriate FDR-specific awards

❑ Prepare AAR comments and document lessons learned

7.4.3 J-2 Intelligence

J-2 responsibilities in support of FDR operations include gathering information on weather, roads, environmental factors, and incident information that are necessary for effective FDR operations. Typical information requirements include: situational awareness assessments, damage assessment, evacuation monitoring, search and rescue tracking, HAZMAT incident tracking/assessment, and hydrographic surveys.

> **NOTE:** *In a FDR environment, do not use the terms "Intelligence, Surveillance, and Reconnaissance (ISR)" or "Intelligence Preparation of the Battlefield (IPB)." It communicates the wrong message to the media and the Affected States/NGO/IO organizations. Recommend use of "Information Operations."*

7.4.3.1 J-2 Preparation for Deployment

❑ Conduct intelligence assessment

❑ Develop situation assessment brief

❏ Develop situation overview to determine scale of humanitarian crisis
❏ Develop geographic overview
  o Determine potential APOD/SPOD locations and characteristics
❏ Obtain weather overview from METOC and determine potential impacts to operations
❏ Obtain geo-spatial products (maps, charts)
❏ Provide Geospatial Intelligence (GEOINT) [commercial imagery, National Technical Means (NTM) and Geospatial] to staff and subordinate HQs
❏ Consult Foreign Disclosure Officer for releasability and eligibility procedures regarding the provision of imagery to NGOs, IGOs, Affected State, and other relief providers
❏ Monitor disaster actions via websites (see Appendix D for additional sites):
  o USAID http://www.usaid.gov/our_work/humanitarian_assistance/disaster_assistance/
  o Relief Web http://www.reliefweb.int
  o One Response http://www.oneresponse.int
  o UN OCHA http://ochaonline.un.org
  o USGS http://www.usgs.gov
  o Center for International Disaster Information http://www.cidi.org
  o Reuters AlertNet http://www.alertnet.org
  o UN Humanitarian Information Center http://www.humanitarianinfo.org
❏ Monitor higher HQ disaster response activities
❏ Determine other participants involved in crisis response
❏ Provide or obtain infrastructure assessments to include lines of communication (LOC) and APODs/SPODs
❏ Provide area assessment of affected region
❏ Collect data from Surgeon and Country Team sources to include:
  o Population-at-risk estimates
  o Casualty estimates
  o Fatality estimates
  o Displaced persons estimates
  o Identify Affected State response mechanism structure
  o Identify response capacity and shortfalls for:
    ▪ Facilities

- Personnel
- Materiel

❑ Assess potential threats to US forces and critical infrastructure in the region
❑ Be prepared to provide a representative for the advance party
❑ Submit information collection requirements to other USG agencies

7.4.3.2 J-2 Deployment

❑ Participate in staff planning efforts
❑ Establish information gathering methodologies
❑ Establish contact with J-2 counterparts
❑ Gain approval to release commercial imagery to NGO, Affected State, and other relief providers

> **NOTE:** *Each traditional ISR system has unique releasability guidance. All imagery, to include photos taken by hand-held cameras, must be approved prior to release (consult with J-2 and Foreign Disclosure Officer).*

❑ In coordination with the Engineer, provide infrastructure assessments (to include LOCs and APODs/SPODs) to J-3/J-4
❑ In coordination with the Surgeon, review medical information on local and regional health threats; ensure that medical threat information is reported to appropriate authorities
❑ Maintain visibility on the international humanitarian community and other stakeholders actions
❑ Participate in Joint Planning Group to assist development of transition plan
❑ Monitor threat to deployed forces
❑ Identify any HAZMAT concerns in the AO
❑ Develop procedure for communicating HAZMAT issues
❑ Plan for reacting to escalating hazards such as fires, chemical spills, ruptured pipelines, and civil disturbance

7.4.3.3 J-2 FDR Operations

❑ Participate in staff planning efforts
❑ Monitor and assess threat to deployed forces and critical infrastructure, and mission
❑ In coordination with J-3/5/7, protect the force by:
   o Conducting all-hazards threat assessment
   o Implementing baseline Force Protection Condition (FPCON) and other directed force protection measures

- o Prescribing appropriate Personal Protective Equipment (PPE)
- o Directing security measures to mitigate risk
- ❏ Ensure risk assessment is updated
- ❏ Receive and process RFIs
- ❏ Provide locations and detailed situational information about HAZMAT to HAZMAT response teams
- ❏ Assist in locating hazards or potential threats
- ❏ Assist in determining numbers and locations of internally displaced persons (IDP)
- ❏ Determine need for Military Information Support Team (MIST)
- ❏ Conduct damage assessments and determine impacts
- ❏ Determine status of LOCs:
  - o Major roads
  - o Railroads
  - o Waterways
  - o Ports
  - o Airports

7.4.3.4 J-2 Transition

- ❏ Collect and consolidate all journals, reports, records, and notes for input to the AAR and subsequent filing in accordance with DOD guidance
- ❏ Review all journal entries and verify availability of supporting documents

7.4.3.5 J-2 Redeployment

- ❏ Determine and execute close-out activities
- ❏ Safeguard and transfer sensitive information and imagery collected during operation in accordance with DOD guidance
- ❏ Prepare AAR comments and document lessons learned

7.4.4 Foreign Disclosure Officer

Foreign Disclosure Officers (FDO) and Designated Disclosure Authority (DDA) are the release authorities for all Controlled Unclassified Information (CUI) and Classified Military Information (CMI). The PAO is the release authority for all unclassified material intended for public release. The FDO will keep logs and records of all disclosures in accordance with national policy and DOD guidance. To ensure operational information is consistently and accurately disseminated, JTF Commanders should maximize employment of the FDO and PAO.

7.4.5 J-3/5/7 Plans & Operations

The expeditionary nature of FDR operations requires great flexibility in both planning and execution. The JTF operations directorate (J-3) assists the commander in directing and controlling operations. US forces will be operating in support of the US Embassy, USAID/OFDA, the Affected State, and in coordination with NGO and IGOs. The Plans and Operations Sections should consider the MDRO Disaster Management Plan for the Affected State; USAID/OFDA Memoranda of Agreement with other USG agencies; and the GCCs CONPLAN before issuing operational guidance.

7.4.5.1 J-3/5/7 Preparation for Deployment

- ❑ Review FDR training opportunities listed in Appendix E
- ❑ Review higher HQ CONOPs
- ❑ Develop Commander's estimate
- ❑ Obtain copy of CJCS Execute Order (EXORD) and implement it
- ❑ In coordination with J-4/8 capture all costs associated with FDR operations (see Appendix C-9)
- ❑ Establish liaison to higher HQ staff or Embassy DATT in Affected State as appropriate
- ❑ Develop request for forces (RFF) requirements
- ❑ Maintain current situational awareness and common operating picture
- ❑ Provide staff with necessary mission analysis information
- ❑ Ensure theater entry requirements have been met by all deploying personnel
- ❑ Develop Operations Order (OPORD)
- ❑ Update staff estimate for OPORD
- ❑ Brief courses of action to Commander for approval
- ❑ Determine potential response options to include forces available, locations, and approximate response times
- ❑ Determine composition of forces, to include:
  - o Contingency Response Forces
  - o Humanitarian Assistance Survey Team (HAST)
  - o Medical Response Assets
  - o Ground and Aerial Transportation units
  - o Engineer to include Forward Engineer Support Team (FEST)
  - o Water purification
  - o Ground Maneuver, Naval, and Air Forces
  - o Maritime Prepositioning Ship (MPS) Assets
  - o Identify Military Information Support Operations (MISO) personnel requirements
  - o Civil Affairs Teams (Army, Navy, USMC)

- ❑ Begin validation of Time-Phased Force Deployment Data (TPFDD)
- ❑ Register for USAID/OFDA fact sheet electronic distribution (http://www.usaid.gov)
- ❑ Deploy assessment teams as directed
- ❑ Identify if communications, logistics, transportation, or other augmentation of forces is required and submit request for forces to higher HQ
- ❑ Conduct assessments in accordance with Section 5.3.3
- ❑ Identify other forces on station; obtain mission sets and areas of responsibility
- ❑ Disseminate Joint Staff SITREP formats and reporting requirements to subordinate elements
- ❑ Brief Courses of Action to Commander for approval
- ❑ In coordination with the Comptroller/Resource Manager, determine and distribute funding guidance
- ❑ Capture all expenditures in support of disaster relief operations are tracked
- ❑ Review previous lessons learned
- ❑ Review Theater Security Cooperation Plan (TSCP)

7.4.5.2 J-3/5/7 Deployment

- ❑ Lead and participate in staff planning
- ❑ Establish a JOC
- ❑ Establish Joint Personnel Recovery Center (JPRC)
- ❑ Establish suspense for daily SITREP
- ❑ Be prepared to accept LNOs
- ❑ Request clearance information and provide badges for LNOs
- ❑ In coordination with J-1/J-4/J-8/Comptroller/Resource Manager begin capturing and reporting cost associated with FDR operations
- ❑ Continue information exchange with USAID/OFDA
- ❑ Post information on International Humanitarian Community (IHC) activity via GCC JOC Crisis Action web page
- ❑ Lead staff planning efforts in development of transition plan
- ❑ Maintain direct coordination and liaison with appropriate organizations/agencies as required
- ❑ Monitor Force Protection Posture
- ❑ In coordination with COS, J-9, and PAO refine Strategic Communication Guidance as necessary
- ❑ Identify requirements for potential follow-on forces
- ❑ Monitor ongoing operations
- ❑ Provide daily SITREP to higher HQ

7.4.5.3 J-3/5/7 FDR Operations

- ❑ Assess Force Protection posture
- ❑ In coordination with J-2, protect the force by:
    - o Conducting all-hazards threat assessment
    - o Implementing baseline Force Protection Condition (FPCON) and other directed force protection measures
    - o Prescribing appropriate Personal Protective Equipment (PPE)
    - o Directing security measures to mitigate risk
- ❑ Monitor operations, disposition of forces, and mission readiness status
- ❑ Lead Mission Tasking Matrix (MiTaM) Working Group
- ❑ Develop and manage MiTaM tracking log
- ❑ Request J-9 identify Civil Affairs (CA) assets deployed within Affected State
- ❑ Track RFFs
- ❑ Prepare mission orders
- ❑ Develop and manage air tasking order
- ❑ Lead staff planning
- ❑ Monitor threat to deployed forces and critical infrastructure
- ❑ Issue OPORD for Phase IV transition and Phase V redeployment
- ❑ Be prepared to develop and implement coordination and control procedures in support of UN Cluster operations
- ❑ Provide daily SITREP to GCC
- ❑ Validate TPFDD for redeployment
- ❑ Begin preparation for AAR

7.4.5.4 J-3/5/7 Transition

- ❑ In coordination with J-4, be prepared to conduct customs and agricultural inspections
- ❑ Monitor handoff of missions and assist as necessary
- ❑ Collect and consolidate all journals, reports, records and notes for input to the AAR and subsequent filing in accordance with DOD guidance
- ❑ Transfer information and responsibilities to Affected State authorities or international relief missions
- ❑ Develop and issue Redeployment Order

7.4.5.5 J-3/5/7 Redeployment

- ❑ Plan and conduct the AAR and capture lessons learned
- ❑ Integrate external organizations and agencies input into the AAR
- ❑ Monitor threat to redeploying forces and critical infrastructure

7-17

❑ Set post-event Force Protection Condition (FPCON) and travel restrictions

❑ Compile entire staff AAR comments and lessons learned and forward to appropriate agencies

7.4.6 Air Component Coordination Element

The Joint Force Air Component Commander (JFACC) may establish one or more Air Component Coordination Element (ACCE) teams to integrate air operations within the joint task force. The ACCE does not replace or circumvent normal theater air tasking request mechanisms. ACCE teams will provide the following support to the JTF:

- Reach-back to JFACC air tasking processes
- Knowledge of JFACC capabilities and limitations
- Assist in communicating JTF support requirements to JFACC
- Facilitates information flow between the JTF component commands, the JTF, and the JFACC

For more information on the JFACC and the air tasking order process, see Section 9.2.8.2.

7.4.7 J-4/8 Logistics & Resource Management

Logistics in FDR operations are complex. The issues of delivering equipment and supplies into disaster decimated and potentially austere environments; the chronic issue of the uncontrolled "pushing" of logistical assets into underprepared distribution centers; and the lack of a single, central control authority combine with routine frequency to complicate FDR logistical operations.

J-4 responsibilities and authority must be clearly delineated to ensure uninterrupted sustainment of ongoing and future operations. The often austere environment in which JTFs will operate may require coordination of common logistic efforts necessary for JTF mission accomplishment. Implications of logistic support must be anticipated and the J-4 must be prepared to integrate these capabilities into the overall support plan, as necessary.

The JTF J-4 will play a key role in coordinating the logistical management of US military assets and will be looked upon by US inter-agency partners, NGO/IGO organizations, and UN cluster leads to help define and coordinate complex logistical operations. While some of those organizations and agencies may not require direct support from the JTF, they will compete for space at ports, airfields, transportation nodes, and local commercial support capacity.

NOTE: *To ensure FDR missions do not negatively affect DOD operational and maintenance budgets, it is critical that accurate accounting of expenses takes place. This will facilitate timely DOD reimbursements from the Department of State/USAID.*

Bills and vouchers shall be processed by the Military Departments and forwarded to the DOD Coordinator for Foreign Disaster Relief. The DOD Coordinator will arrange to have them aggregated and forwarded to the Department of State for payment, per DODD 5100.46.

When preparing billings for reimbursement of costs incurred, the following guidelines apply:

- Materials, supplies, and equipment from stock will be priced at standard prices used for issue to DOD activities, plus applicable accessorial costs for packing, crating, handling, and transportation
- Materials, supplies and equipment determined to be excess to the DOD will be available for transfer to the Department of State without reimbursement, in accordance with established DOD policies. Accessorial charges for packing, crating, handling, and transportation will be added where applicable
- Air and ocean transportation services performed by the Air Mobility Command (AMC) and the Military Sealift Command (MSC) will be priced, where applicable, at current tariff rates for DOD Components. Where tariff rates are not applicable, air transportation, whether provided by AMC or other aircraft, will be priced at the "Common-User Flying-Hour," rate for each type of aircraft involved and ocean transportation provided by MSC will be priced at "Ship Per Diem Rates"
- Services furnished by activities under DOD Industrial Funds other than AMC and MSC will be priced to recover direct and indirect costs applicable to reimbursements for services rendered to other Department of Defense activities
- Personal services furnished will be priced at rates to recover overtime of civilian personnel
- All other services furnished, not specifically covered above, shall be priced on a mutually agreeable basis and, if feasible, such prices shall be established prior to the services being furnished. Prices for such services shall be at the same rates that the Department of

7-19

Defense would charge other Government Agencies for similar or like services, if such rates are available; otherwise the basis of pricing will be to effect full reimbursement to the Department of Defense appropriations for "out-of-pocket" expenses

NOTE: *Effective logistics support operations in a FDR environment are critical. The DOD is in direct support of the Affected State, the US Embassy, and USAID/OFDA. In addition, DOD units may be required to logistically support civilian organizations as well as provide for individual unit's needs. Knowing the logistics plans of the support agencies will create a strong working relationship.*

7.4.7.1 J-4/8 Preparation for Deployment

- ❑ Review FDR training opportunities listed in Appendix E
- ❑ Obtain CJCS EXORD
- ❑ Determine Overseas Humanitarian Disaster and Civic Aid (OHDACA) appropriation funding available for FDR mission
- ❑ Coordinate with Staff Judge Advocate (SJA) on legal authorities prior to utilization of non-OHDACA funding (see Appendix A-6 for other funding sources)
- ❑ Obtain funding guidance from higher HQs that prescribes reimbursement and cost reporting requirements from GCC J-8
- ❑ In coordination with J-3/5/7 capture all costs associated with FDR operations (see Appendix C-9)
- ❑ Be prepared to deploy a logistics representative with the advance party
- ❑ Request verification of CJCS Project Code for FDR operations from GCC J-4
- ❑ Review Theater Security Cooperation Plan (TSCP)
- ❑ Obtain copy of JS/OSD approval letter for lift of non-DOD material and personnel
- ❑ Define logistics manning requirements for assessment teams
- ❑ Identify potential sources of FDR supplies
- ❑ Estimate how a reduced infrastructure will impact supply distribution and standard consumption rates
- ❑ Review Acquisition Cross-Service Agreements (ACSA), Mutual Logistics and Services Agreements (MLSA), and/or Cooperative Security Location (CSL) statuses
- ❑ Evaluate assessment data on APOD and SPOD to facilitate movement of equipment, supplies, and personnel
- ❑ Establish single collection point for assessments/requirements

❏ Be prepared to provide personnel to support Joint Deployment & Distribution Operations Center (JDDOC)
❏ Establish daily LOGSTAT suspense
❏ Estimate logistics support requirements for each class of supply for theater and other supported elements
  o Class I – Food—when and where units will be fed; sources of potable drinking water
  o Class II – Durable supplies and equipment
  o Class III – Bulk and packaged POL products
  o Class IV – Barrier materials—for safety and security of unit personnel, on-going weather phenomena, and impacts on personnel comfort (may need tarps and plywood flooring)
  o Class V – Ammunition—resupply usually not required FDR operations; plan to deploy with sufficient quantities to protect property and equipment or for show of force
  o Class VI – Personal items—where and how personnel will obtain basics (toothpaste, shaving cream, deodorant, stamps, envelops, gloves, undergarments, t-shirts); where personnel will wash clothes and shower
  o Class VII – Major end items—for maintenance service and repair capabilities that exceed unit's organic repair capabilities
  o Class VIII – Medical supplies, equipment, and biomedical repair parts
  o Class IX – Repair parts—deploy with standard shop stock/ bench stock and Authorized Stockage List (ASL) items to support minimum of 30 DOS
❏ Be prepared to attend and participate in Logistics Cluster meetings with USAID/OFDA

> **NOTE:** *Remind personnel that power sources in an Affected State may be 220 versus 110 volts. Suggest taking power converters, batteries, extension cords, multi-plug devices, charging devices, power generation equipment as well as parts, and fuel for essential communication equipment.*

7.4.7.2 J-4/8 Deployment
❏ Track operational cost in accordance with higher HQs and comptroller guidance (see Appendix C-9)

7-21

- ❑ Validate logistics support requirement estimates for each class of supply
- ❑ Determine potential supply providers and transportation requirements
- ❑ Determine aviation and ground transportation logistics and sustainment requirements
- ❑ Manage maintenance and supply actions
- ❑ Determine sources for petroleum, oil, and lubricants (POL)
- ❑ Plan for military sustainment requirements (e.g., shelter, power, mess, rations, water, bath, laundry, etc.)
- ❑ Determine if Morale, Welfare and Recreation (MWR) phones are available (availability and number vary by incident)
- ❑ Determine military versus civilian supplied items
- ❑ Determine military vehicle restrictions
- ❑ Plan for weapons storage and guard force, if necessary
- ❑ Provide logistics updates to supporting response partners
- ❑ Be prepared to participate in Logistics Cluster meetings
- ❑ Establish or coordinate with JDDOC
- ❑ Conduct total asset visibility (TAV)
- ❑ Coordinate channel flights with GCC JDDOC and TRANSCOM
- ❑ Conduct Logistics Coordination Board (LCB)
- ❑ Compile daily LOGSTAT report and submit to higher HQ
- ❑ Participate in a Joint Planning Group (JPG) to assist development of transition plan

7.4.7.3 J-4/8 FDR Operations

- ❑ Continue tracking detailed mission costs and financial expenditures; *Keep receipts.* The following is a partial list of items to include (see "Resource Tracking" in Appendix C-9):
    - o Record missions performed, with particular attention to aviation mission support and other high dollar missions
    - o Rosters of personnel involved
    - o Travel and per diem (military and civil service)
    - o Lodging cost
    - o Transportation cost (car and bus rentals, chartered aircraft, and fuel)
    - o All contracting costs
    - o Equipment provided or operated (estimated hourly cost)
    - o Materiel provided from regular stock (all classes of supply)
    - o Laundry expenses
    - o All classes of supply expended

❑ Establish policies and procedures for military personnel transportation

❑ Determine specialized equipment requirements (e.g., cold weather, medical supplies)

❑ Plan for maintenance requirements

❑ Coordinate with Affected State for possible use of state maintenance facilities for equipment and vehicles

❑ Participate in MiTaM Working Group

❑ Identify contracting options and requirements

❑ Establish or coordinate with JDDOC

❑ Conduct total asset visibility (TAV)

❑ Monitor channel flights

❑ Monitor APOD/SPOD condition

❑ Conduct and participate in Logistics Coordination Board (LCB)

❑ Compile daily LOGSTAT report and submit to higher HQ

❑ Provide daily logistics situation report (LOG SITREP) to Higher HQ J-4

❑ Confirm locations and sources to purchase parts, POL, and supplies

❑ Procure, transfer, and distribute food, water, and supplies

❑ Submit statements of work, requirements, and determinations through the supporting contracting team

❑ Establish controls for use of government purchase cards

❑ Coordinate with JTF Surgeon, medical officer, or team for proper disposal of biohazardous material

❑ Provide transportation support to response assets

❑ Plan vehicle recovery and extraction

❑ Confirm waste removal plans

❑ Identify and contract with local vendors for media reproduction support (e.g., high-speed, large format printing)

❑ Participate in staff planning and OPORD development

7.4.7.4 J-4/8 Transition

❑ In coordination with J-3/5/7, be prepared to conduct customs and agricultural inspections

❑ Assign logistics officer to clear remaining logistics issues

❑ Close out all remaining contracts

❑ Submit reports and request reimbursement

❑ In demobilization procedures, include clearing base camp of equipment provided by civilian authority (including COTS service agreements) and disposing of COTS equipment

❑ Close out with JDDOC

❑ Conduct TAV

❑ Provide daily logistics situation report (LOG SITREP)

7.4.7.5 J-4/8 Redeployment

❑ In coordination with J-3 and higher HQs compile mission execution data including but not limited to total man-hours used, number and type of equipment used, fuel usage, maintenance performed and equipment lost, damaged or destroyed
❑ Complete accounting and turn-in of any unused supplies
❑ Conduct TAV
❑ Provide final daily logistics situation report (LOG SITREP)
❑ Prepare AAR comments and document lessons learned

7.4.8 Engineer

The JTF staff Engineer develops and coordinates engineering requirements for the JTF Commander. The staff Engineer is the primary engineering interface between all Joint Task Force elements, including non-military assets from DOS, USAID, UN, Affected States, NGOs, IGOs, contingency contractors, and theater construction organizations. The engineer ensures efforts are synchronized and provides the JTF Commander with an Engineering Common Operating Picture. Engineer planners should be included in the planning process to ensure integration of the Engineer Support Plan (ESP) into the JTF Commander intent and strategy.

It is essential to establish an engineer liaison in the CMOC to coordinate and execute engineering support with USAID, NGOs and other organizations. The JTF staff Engineer should be cognizant of NGO engineering activities to ensure that DOD efforts are not duplicative and make best use of limited operational resources.

FDR operations can be engineer-intensive. In such cases, JTF commanders may consider establishing an independent Engineer Task Force to manage engineer operations. Such a task force may be formed around an existing engineer command or naval construction regiment. Engineer forces could be placed under OPCON, TACON, or in a supporting role. The engineer assets attached to the subordinate JTF will normally be made up of a mix of engineer assets drawn from the entire force's engineer resources. If the subordinate JTF is to provide a common support capability, it will require a specific delegation of directive authority from the CDR.

The Task Force Engineer will have to coordinate with USG agencies and NGOs conducting engineer activities. NGOs may have unique engineering capabilities that can be leveraged as part of the overall operational effort. Engineer tasks and supporting organizations are listed in Section 11.7.

The level of assistance that can be provided by engineers may vary from limited, highly specialized teams to complete control of all engineer units.

7-24

Specialized teams are used to assess damage or estimate engineering repairs and can assist in unique support such as power supply and distribution; utilities repair work, water purification, and well drilling operations. In large FDR operations, engineer units provide essential general engineering support including facility construction, structural repair, and camp construction for deployed forces.

7.4.8.1 Engineer Preparation for Deployment

❑ The engineer staff must be prepared to conduct/manage the following engineer operations:
- o Engineer assessment of critical infrastructure
- o JTF base camp construction
- o Construction projects in support of IDPs
- o APOD/SPOD assessment repair
  - ▪ airfield damage repair
  - ▪ port repair
  - ▪ vessel salvage
- o Repair of critical services
  - ▪ emergency power/lighting
  - ▪ water purification
  - ▪ water distribution
  - ▪ sewage treatment
  - ▪ sanitation
- o Repair of critical transportation routes
  - ▪ roads
  - ▪ bridges
  - ▪ culverts
- o Debris and obstacle clearing
- o Well repair and well drilling
- o Training – Affected State ministries, local government, local contractors, military or other organizations

❑ Determine Humanitarian Assistance Survey Team (HAST) and Contingency Command Post engineer personnel requirements for 24/7 operations (Joint Manning Document)
- o Validate required credentials (i.e. Electrical, water distribution, sanitation, civil/structural, environmental)
- o Provide team with local government POCs/other US agency POCs names and numbers
- o Recommended equipment:
  - ▪ Camera
  - ▪ Laser measurement tool
  - ▪ Hand-held GPS
  - ▪ Laptop computer

- ❑ Obtain initial damage assessment of designated area of operations-including critical infrastructure-power distribution, water treatment/distribution plants, APOD and SPOD
- ❑ Contact Country Team and military assistance group for a list of key facilities and critical infrastructure
- ❑ Obtain geo-spatial data (i.e. maps, geo-technical – current and historical erosion/flood zones, and infrastructure – road, bridges, and river data)
- ❑ Request Embassy/contracting command element in AOR to identify availability of Class IV materials, engineer related-equipment, suppliers, and contractors
- ❑ Identify USACE/NAVFAC assets in country or AOR
- ❑ Identify possible facilities and real estate for JTF operation sites – base camps, APOD/SPOD operations
- ❑ Validate that engineer units on Prepared To Deploy Orders have capabilities to address preliminary engineer requirements
- ❑ Identify typical disaster-related engineer contracts, synchronize with local contracting office, and bring draft Statements of Work for down range contracting (e.g., APOD and SPOD repairs, base camp construction, debris removal, critical services assessment and repairs)
- ❑ Identify possible Contracting Officer Representative requirements for typical contracts (e.g., APOD and SPOD repairs, base camp construction, debris removal)
- ❑ Provide the J-2 with Geo-spatial resources support
- ❑ Support the Operations Protection Division effort with any Force Protection and civil security construction and repair requirements
  - o Protect FDR forces
  - o Repair prisons and local policing facilities
- ❑ Support the J-6 with any communications infrastructure construction requirements
- ❑ Obtain geo-spatial data i.e. maps, geo-technical and historical data
- ❑ Obtain technical information-construction materials, building codes, availability of Class IV, suppliers, contractors
- ❑ Identify key facilities and key infrastructure
- ❑ Determine HAST personnel requirements to conduct Engineer Assessments
- ❑ Identify potential service contracts and request legal review from SJA

❑ Identify LNOs
- o USAID-DART
- o CMOC
- o Military Group (MILGROUP)
- o Affected State and Assisting States Civil and Military Engineers
- o Identify NGO/IGO Engineers

7.4.8.2 Engineer Deployment

❑ Synchronize engineer effort with USAID/OFDA and Country Team to ensure priority of effort

❑ Identify and recommend engineering related Commander's Critical Information Requirements (CCIR) and provide updates

❑ Support J-4 in engineer related sustainment requirements

❑ Determine Class IV supply providers, locations, and transportation availability

❑ Develop the Engineer Support Plan that meets USAID/OFDA and Country Team priorities

❑ Develop and coordinate tasks for component engineer forces

❑ Conduct engineering assessments

❑ Continue to obtain technical information-construction materials, building codes, availability of Class IV, suppliers, contractors

❑ Determine if engineering assessments have been conducted on bridges, roads, APODs and SPODs to support deploying forces

❑ Obtain assessments of key facilities and key infrastructure

7.4.8.3 Engineer FDR Operations

❑ Manage and synchronize the repair of APODs and SPOD

❑ Manage construction of JTF bases of operations

❑ Conduct Engineer Critical Infrastructure Assessments-bridges, roads, and critical facilities

❑ Receive guidance from the CMOC meetings and report actions to Joint Civil-Military Engineering Board (JCMEB) if established

❑ Participate in Joint Acquisition Review Board for all engineer related requirements

❑ Develop all engineer contracting requirements-write Statements of Work, Bills of Materials, and all information required to establish contracts

❑ Manage all DOD engineer efforts in order to provide the JTF Commander a complete Engineer Contingency Operations Picture

❑ Identify tasks to be executed by US Army Corps of Engineer (USACE), Naval Facilities Command (NAVFAC), or other subject matter expert teams; such as Forward Engineer Support Teams

(FEST); Contingency Real Estate Support Teams (see Chapter 11 for description of engineer units)

❑ Coordinate and facilitate the Joint Facilities Utilization Board, Joint Civil Military Engineer Board, and Joint Environmental Management Board

❑ Screen, validate, and prioritize all engineering projects and mission assignments

❑ Prepare logistic reports on engineer resources using the Joint Operation Planning and Execution System

❑ Plan and coordinate the distribution of construction and barrier materials based on established priorities

❑ Establish the statement-of-work, development of contracts, and employment of services

❑ Plan and provide guidance for environmental considerations (including baseline surveys and follow-on assessments) that impact joint operations

❑ Train Affected State personnel to conduct construction or other contracted engineer projects and to operate, maintain new and established infrastructure systems

7.4.8.4 Engineer Transition

❑ Ensure all existing contracts are closed out or transitioned to the Country Team, USAID, or the Affected State

❑ Ensure all real estate leases are closed out or extended as needed

❑ Close out all USACE/NAVFAC projects or transition management/oversight to the Service Component Command or responsible GCC

❑ Identify Base Camp closure requirements (include environmental, as required)

❑ Support J-4 in Class IV redistribution plan for materials that have not been utilized/turn in

❑ Finalize hand off of Engineer operations to trained Affected State personnel the operation and maintain established infrastructure

❑ Assist in planning redeployment of organic engineer resources

❑ Receive *as-builts* from all construction contractors/units

7.4.8.5 Engineer Redeployment

❑ Finalize *as-builts* of all construction/repair contracts and provide to Affected State

❑ Prepare AAR comments and document lessons learned

7.4.9 Contingency Contracting

Contract support is delivered to the joint force through a process comprised of five key tasks: planning; requirements determination; contract development; contract execution; and contract closeout. The product of planning is a contract support integration plan that defines key contract support integration capabilities. Contingency acquisition, the process of acquiring supplies, services, and construction in support of the joint operations, begins at the point when a requiring activity identifies a specific requirement, defines the proper funding support, contract award, and administration requirement to satisfy activity needs.

> *JP 4-10 Joint Contingency Contracting Manual and Contingency Contracting Handbook website:*
> *http://www.acq.osd.mil/dpap/ccap/cc/jcchb/intro_contin.html*

7.4.9.1 Contingency Contracting Preparation for Deployment

- ❏ Develop Contract Support Integration Plan/Contract Management Plan, other Operational Contract Support (OCS) guidance
- ❏ Identify staff and subordinate command OCS responsibilities
- ❏ Determine unit OCS related training requirements
- ❏ Prepare for Pre-Deployment Site Survey
- ❏ Finalize habitual contractor support arrangements

7.4.9.2 Contingency Contracting Deployment

- ❏ Complete Phase 1 requirements and be prepared to execute contracts for initial JTF Operations

7.4.9.3 Contingency Contracting FDR Operations

- ❏ Manage all contracts working with the various Contracting Officer Representatives (COR)

7.4.9.4 Contingency Contracting Transition

- ❏ CORs submit close-out forms to ensure projects are completed and contractor's are entitled to final payment

7.4.9.5 Contingency Contracting Redeployment

- ❏ Close out all contracts completed/all contractors paid
- ❏ AAR comments drafted for input

7.4.10 J-6 Command, Control, Communications, and Computer Systems

The JTF J-6 assists the commander in all responsibilities for communications infrastructure, communications-computer networking, communications electronics, information assurance, tactical communications, and interoperability. This includes development and integration of communications system architecture and plans that support the command's operational and strategic requirements, as well as policy and guidance for implementation and integration of interoperable communications system support to exercise command in the execution of the mission. The communications system supports a collaborative information environment that assists commanders in conducting detailed, concurrent, and parallel planning.

Internal military communications can be accomplished in accordance with unit standard operating procedures; however, requirements to communicate with the Affected State, US Embassy, USAID, NGO, and IGOs may present challenges. Depending upon the situation, units may be tasked to provide communications for non-DOD agencies/organizations. Expect that military communications equipment may not be compatible with civilian equipment.

7.4.10.1 J-6 Preparation for Deployment

- ❑ In coordination with higher HQ information manager establish Information Management/Knowledge Management website
- ❑ Post NGO/IGO-USAID/OFDA information on crisis action web page
- ❑ Be prepared to deploy communications representative with the advance party
- ❑ Establish JOC and TOC Crisis Action web pages (classified and unclassified) and web portal as required
- ❑ Reviews POC list for the specific country or region points of contact (i.e. US Embassy MDRO, DATT, foreign military representatives, USAID/OFDA, etc) and post on crisis webpage
- ❑ Condense existing POC list for the specific country or region (embassy, foreign military, USAID/OFDA, etc) and post on the JOC crisis webpage
- ❑ Ensure initial communications capabilities are self-sufficient
- ❑ If possible, coordinate interoperability with local authorities
- ❑ Do not send equipment without operators, essential repair parts, manuals, tools, and initial fuel and power generation requirements (items may be difficult to obtain in the affected area)
- ❑ Determine if military will be required to provide communications availability to other agencies. If yes, then:

- o   Determine configuration requirements
- o   Procure equipment (all sources—military, other federal agencies, and civilian)
- o   Establish data and voice protocols
- o   Keep good records for funding reimbursement

❑ Plan for all means of communications and purchasing of additional communication devices/services:
- o   Telephone (satellite, cellular, or landline)
- o   Radio (military, maritime, and civilian, in all bandwidths)
- o   Non-classified Internet Protocol Router Network (NIPRNET)
- o   Secret Internet Protocol Router Network (SIPRNET)
- o   Video equipment and video teleconferencing
- o   Satellite-based commercial internet systems

> **NOTE:** *Remind personnel that power sources in an Affected State may be 220 versus 110 volts. Suggest taking power converters, batteries, extension cords, multi-plug devices, charging devices, power generation equipment as well as parts, and fuel for essential communication equipment.*

7.4.10.2 J-6 Deployment

❑ Issue communications plan
❑ Establish and maintain communications architecture (Internet, telephone landline, and cell phone network)
❑ Publish military phonebook and acquire important civilian point of contact (POC) listings
❑ Determine initial communications package and potential expansion requirements
❑ Assist development of transition plan
❑ Establish method to collect lessons learned input
❑ Track message traffic with higher HQ and pass to appropriate staff elements
❑ Facilitate frequency management de-confliction for deploying forces
❑ Identify and provide links to products from previous FDR operations
❑ Orient the Common Operating Picture to the area of interest
❑ Build newsgroups to support TPFDD flow in accordance with JOPES

- ❏ Determine need for Information Operations planner support for participating units
- ❏ Develop and release Operational Security guidance to participating units
- ❏ Establish lessons learned input page for JOC Crisis webpage

7.4.10.3 J-6 FDR Operations

- ❏ Participate in staff planning process
- ❏ Ensure all communications and information technology personnel are qualified
- ❏ Maintain communications architecture as required for Joint Operations Center (JOC)/Tactical Operations Center (TOC)/Civil Military Operations Center (CMOC)
- ❏ Monitor status of satellite phones, cell phones, and Internet connectivity
- ❏ Provide communications equipment (cell phones, radios, base sets, etc.) to LNOs
- ❏ Participate in MiTaM Working Group
- ❏ Prepare daily update brief slide

7.4.10.4 J-6 Transition

- ❏ Ensure accountability for all communications equipment loaned or borrowed
- ❏ Cancel COTS service agreements, and dispose (by standard operating procedures) of COTS equipment

7.4.10.5 J-6 Redeployment

- ❏ Prepare AAR comments and document lessons learned

7.4.11 J-9 Civil Affairs/Interagency/International Cooperation

The J-9 is the principal advisor to the commander and staff on Civil Affairs Operations (CAO). He advises on the capabilities, allocation, and employment of subordinate Civil Affairs (CA) units and provides specific country information. The J-9 counterpart at the Embassy is the Cultural Affairs Officer.

The J-9 staff develops the CA Operations annex to OPLANs and CONPLANs. The J-9 coordinates with supporting CA forces and CMOC to conduct Interagency (IA) collaborative planning/coordination and integration of non-military stakeholders with the staff. The J-9 ensures the timely update of the Civil Affairs portion of the Common Operational Picture (COP) through the Civil Information Management (CIM) process. The J-9 prepares and maintains the CAO running estimate and advises the commander on legal and moral obligations incurred from the long- and short-term effects (economic, environmental, and health) of military operations on civilian populations.

CA forces offer unique capabilities that not only enhance the military mission but also ultimately advance US interests. Properly employed CA forces help shape the environment and set the conditions for transition operations. CA operations involve the interaction of military forces with the civilian populace to facilitate military operations and consolidate operational objectives. A supportive civilian population can provide resources and information that facilitate friendly operations. It can provide a positive climate for the military and for the nation to pursue diplomatic activities that achieve foreign policy objectives.

7.4.11.1 J-9 Preparation for Deployment

- ❑ Participate in preparation and review of contingency plans
- ❑ In coordination with COS, J-3, and PAO refine Strategic Communication Guidance
- ❑ Track deployed civil affairs forces through daily CMO SITREP
- ❑ Establish link and brief J-staff on civil affairs forces
- ❑ Monitor all FDR operations for compliance with applicable laws, agreements, treaties, and contracts
- ❑ Review guidance from the GCC regarding FDR operations and disaster relief plans
- ❑ Incorporate FDR assessment and training into TSCPs
- ❑ Gather all available information about the civil environment in the Affected State both before and after the disaster
- ❑ Make contact with any UN OCHA element in the Affected State through appropriate channels
- ❑ Track international humanitarian community actions in progress and those being initiated by UN, IGOs, and NGOs
- ❑ Identify participating NGO organizations/agencies
- ❑ Determine NGO agencies functions, operational and strategic objectives
- ❑ Prepare a matrix that lists organizations/agencies involved and agency's functions, operational and strategic objectives
- ❑ Provide CMO planner augmentation to assessment teams
- ❑ Contact component CMO representatives for participation in the Operational Planning Team (OPT)
- ❑ Identify CA assets already operating in Affected State
- ❑ Determine CA force requirements and assist J-3 in preparing RFF for deployment
- ❑ Monitor disaster actions via websites (see Appendix D for additional sites):

- USAID
  http://www.usaid.gov/our_work/humanitarian_assistance/disaster_assistance/
- Relief Web http://www.reliefweb.int
- One Response http://www.oneresponse.int
- UN OCHA http://ochaonline.un.org
- USGS http://www.usgs.gov
- Center for International Disaster Information
  http://www.cidi.org
- Reuters AlertNet http://www.alertnet.org
- UN Humanitarian Information Center
  http://www.humanitarianinfo.org

7.4.11.2 J-9 Deployment

❑ Assess the environment in which US forces will conduct FDR operations, including:
  - Political situation
  - Physical boundaries of the area
  - Potential threat to forces
  - Global visibility of the situation
  - Media interest
  - Climate for FDR operations
  - Assess the Affected State's ability to manage FDR

❑ Be prepared to establish a CMOC to coordinate and synchronize efforts with the interagency, IGOs, and NGOs
❑ Identify shortfalls in Affected State plans and resources
❑ Identify Affected State disaster resources, including various government agencies, military units, IGOs and NGOs, in the AOR, and establish contact and working relationships as appropriate
❑ Assess the impact of FDR operations on the populace
❑ Develop plans and strategies for long-range mitigation of political, economic, legal, social, and military issues associated with FDR operations
❑ Conduct cultural awareness orientations

7.4.11.3 J-9 FDR Operations

❑ Assess the environment in which US forces are conducting FDR operations, including:
  - Political situation
  - Physical boundaries of the area
  - Potential threat to forces
  - Global visibility of the situation

- o Media interest
- o Climate for FDR operations
- o Confirm and validate the Affected State's ability to manage disaster relief
- ❑ Identify shortfalls in Affected State plans and resources
- ❑ Identify resources, including various government agencies, military units, IGOs, and NGOs in the theater of operations, and build working relationships
- ❑ Assess, monitor, and report the impact of FDR operations on the populace
- ❑ Develop plans and strategies for long-range mitigation of political, economic, legal, social, and military issues associated with FDR operations
- ❑ Assist other staff elements in the development of transition plans
- ❑ Coordinate with civil authorities to obtain relevant information
- ❑ Maintain visibility on NGO and IGO organizations and agencies and their operations for coordination and cooperation requirements
- ❑ Establish a CMOC or "Information Center" to facilitate sharing of information and coordination of activities between Affected State government, NGOs, IGOs, and the JTF
- ❑ Work closely with USAID/OFDA staff and facilitate USAID/OFDA activities
- ❑ Continue information exchange with USAID mission in country
- ❑ Participate in MiTaM Working Group
- ❑ Coordinate with international humanitarian community for transition of responsibility
- ❑ Maintain direct coordination and liaison with appropriate organizations and agencies
- ❑ Update and revise the NGO and IGO matrix with functions and strategic and operational objectives of organizations and agencies

7.4.11.4 J-9 Transition

- ❑ Maintain direct coordination and liaison with the Affected State and NGO and IGO organizations and agencies
- ❑ Transfer information and responsibilities for FDR operations to Affected State authorities or appropriate IGO, NGO. This may include:
  - o Planning and scheduling meetings to thank military, civilian and interagency participants and supporters
  - o Planning a transition ceremony
- ❑ End all administrative aspects of the mission and create a transition plan that includes:

   o Complete or halt all ongoing projects and programs
   o Pay all fees, rents, and contract amounts owed
   o Close all logs, complete reports and secure all data
   o Secure or destroy all classified material related to Civil Military Operations
   o Reconcile operational funds with Finance Officer
   o Prepare and submit Special Operations Debrief and Retrieval System input

❑ Turn leased or utilized facilities over to appropriate Affected State authorities or appropriate IGO, NGO after you have:
   o Cleaned the facilities
   o Ensured the facilities are returned in the same or better condition as when procured
   o Met any other agreed upon or contractual obligations
   o Scheduled an inspection or orientation tour

7.4.11.5 J-9 Redeployment

❑ Prepare AAR comments and document lessons learned

7.5 Special Staff

Personal and special staff members perform duties as prescribed by the JTF Commander (Refer to Joint Pub 3-29, Appendix A, Commander, Joint Task Force Personnel, and Special Staffs, for a more information on the roles and responsibilities of these staff elements). Special Staff may include (but is not limited to) Public Affairs Officer (PAO), Staff Judge Advocate (SJA), Chaplain (CH), Surgeon, Inspector General (IG), Meteorological and Oceanographic Officer (METOC), Provost Marshal (PM), comptroller, safety officer and political advisor.

7.5.1 Political Advisor

The Political Advisor (POLAD) (may also be referred to as the Foreign Policy Advisor) is a senior State Department officer, detailed as personal advisor to the operational level commander, tasked with providing policy support regarding the diplomatic and political aspects of the commander's responsibilities. To accomplish this tasking, the POLAD must be fully integrated in all activities. The POLAD performs the following duties:

❑ Provides advice to the commander and staff on the political-military aspects of planned and ongoing operations
❑ Analyzes the regional and global political-military environment and the potential impact of planned and ongoing operations
❑ Participates in major command policy and planning activities ensuring political-military issues are fully considered during decision making

❑ Serves as the JTF's primary interlocutor with US Embassy civilian leadership, key decision-makers in the State Department, and with primary officials of other USG agencies in-country

7.5.2 Staff Judge Advocate (Attorney)

The JTF SJA is the principal legal advisor to the JTF Commander. The SJA provides the full spectrum of legal services to the JTF Commander and staff and coordinates with the supported SJA to optimize legal support. As the JTF Commander's personal legal advisor, the SJA normally should have direct access to the JTF Commander. The office must be joint and provide the mix of legal subject matter experts necessary to fully support the JTF. Detailed information on legal aspects of FDR can be found in Appendix A.

7.5.2.1 SJA Preparation for Deployment

❑ Prepare and brief deploying forces at all levels on applicable Affected State considerations to include legal status of deployed forces

❑ Advise the commander on all operational, legal, and fiscal issues

NOTE: When an overseas disaster occurs, the DOS coordinates a memorandum of agreement with the Affected State that includes a blanket release of liability for all US respondents, including licensed and credentialed professions such as health care providers and engineers. A copy of this agreement should be available at the US Embassy/Consulate by the time US forces begin to deploy.

❑ Identify agreements in place with the Affected State
❑ Have copies of the Standing Rules of Engagement (SROE) in Chairman of the Joint Chiefs of Staff (CJCS) Instruction (CJCSI) 3121.01B available for the commander to review; assist Commander in the development of supplemental ROE
❑ Brief deploying forces on applicable Rules of Engagement (ROE) and General Orders (recommend distribution of hardcopy ROE card to personnel)
    o Ensure subsequent ROE briefings are coordinated with ROE Working Group Chair

❑ In coordination with the GCC SJA, develop General Orders for Commander's approval
❑ In coordination with the J1, determine how General Orders, policies, and procedures will apply to deploying civilian personnel

7.5.2.2 SJA Deployment

- ❑ Participate in staff planning process at all levels
- ❑ Contact the SJAs at the GCC and supporting commands to coordinate legal aspects
- ❑ Maintain situational awareness of mission execution and ensure unit activities are consistent with the law
- ❑ Be prepared to deploy required personnel with the advance party
- ❑ Provide status reports, as necessary, to higher headquarters on the following:
  - ○ Criminal incidents
  - ○ Disciplinary/administrative/prosecutorial actions
  - ○ Affected State claims against the US Government

7.5.2.3 SJA FDR Operations

- ❑ Participate in staff planning process
- ❑ Continue to advise commanders and staff on legal matters.
- ❑ Verify that proposed Mission Assignments (MAs) are legally permissible, approved, and executed according to applicable references and restrictions
- ❑ Monitor contract proposals to ensure compliance with fiscal law
- ❑ Participate in MiTaM Working Group

7.5.2.4 SJA Transition

- ❑ Ensure all legal actions are closed or have been referred to appropriate commands before re-deployment
- ❑ Close or refer to the Services all civil/military actions prior to re-deployment
- ❑ Liaise with higher HQs to monitor legal matters

7.5.2.5 SJA Redeployment

- ❑ Prepare AAR comments and document lessons learned

7.5.3 Public Affairs Officer

The Public Affairs Officer (PAO) provides input to the operational planning process, and supports the commander's political, economic, and informational goals, as well as the JTF's assigned objectives. The PAO assesses the impact of military operations in FDR in the context of both the global and military information environments. The PAO assists the JTF Commander in refining and presenting information regarding the command's involvement in FDR operations.

Fostering and furthering good relations abroad is in the best interest of the Department of Defense (DOD Directive 5122.5, "Assistant Secretary of Defense of Public Affairs").

7-38

Affected State leadership determines dissemination protocols regarding emergency information. DOD may assist the appropriate authorities with dissemination of information as long as the instructions are properly attributed.

Service-specific Public Affairs References

Joint Publication (JP) 3-16, Public Affairs

Field Manual (FM) 46-1, Public Affairs Operations

Marine Corps Warfighting Publications (MCWP) 3-33.3, Marine Corps Public Affairs

Secretary of the Navy Instruction (SECNAVINST) 5720.44B, Public Affairs Policy and Regulations

Air Force Instruction (AFI) 35-101, Public Affairs Policies and Procedures

7.5.3.1 PAO Preparation for Deployment

- ❏ Receive and review higher HQ's strategic themes and messages
- ❏ Coordinate with US Embassy Public Affairs Section (PAS) to obtain contact list
- ❏ Coordinate Public Affairs (PA) Guidance with higher headquarters themes and messages
- ❏ Develop consistent messaging and talking points
- ❏ Monitor news media
- ❏ Coordinate with J-1 to establish manning requirements for the Joint Information Center (JIC)
- ❏ Provide military units with current PA guidance prior to entry into the affected area
- ❏ Develop media smart cards based on the PA guidance for issue to all military personnel involved in FDR operations
- ❏ Brief talking points to all personnel prior to deployment

7.5.3.2 PAO Deployment

- ❏ Release Public Affairs Guidance (PAG)
- ❏ Coordinate with US Embassy PAS
- ❏ Designate and provide one or two representatives to the JIC (if established)
- ❏ Brief deploying personnel on media and public engagement policy; distribute media smart cards

❑ Issue guidance to US military personnel regarding personal photography, blogs, social media, and email
❑ Pre-select and train media escorts
❑ Review Strategic Communication (SC) guidance
❑ Develop strategic themes and messages
❑ Facilitate media coverage
❑ Determine Combat Camera (COMCAM) support; draft TASKORD or RFF as necessary
❑ Coordinate with FDO for distribution of imagery and information products via web and media outlets
❑ Brief task force and unit commanders on their roles, responsibilities, and authorities concerning public information requests

Note: Basic Public Affairs Guidance (PAG)
Talking to the media
As a member of the military, you have a great story to tell. Everything you say reflects upon your unit, your Service, and DOD. When speaking with a reporter, everything is on the record.

You may discuss:
- *What you do for the DOD*
- *How you train and prepare to deploy*
- *Where, generally, you are headed—e.g., hurricane in Haiti*

Do not discuss:
- *Personal information*
- *Exact numbers and location of personnel and equipment*
- *Specific aircraft or weapons configurations*
- *Specific force protection measures*
- *Rules of engagement*
- *Classified information*

Do not speculate: If you do not know an answer, say so. If you have questions, contact your local Public Affairs Officer.

7.5.3.3 PAO FDR Operations

- ❑ Monitor information environment (open source and classified); prepare information assessments
- ❑ Review Strategic Communication (SC) guidance
- ❑ Refine strategic themes and messages
- ❑ In conjunction with US Embassy PAS, and JTF and higher HQ, develop transition strategic communications strategy
- ❑ Validate Combat Camera (COMCAM) support requirements
- ❑ Ensure accurate coordination for COMCAM requirements
- ❑ Update PAG as required; obtain release guidance from higher HQ/GCC website; JTF website or US Embassy PAS
- ❑ Continue to coordinate with US Embassy PAS
- ❑ Continue to facilitate media coverage
- ❑ Continue to coordinate with FDO on the distribution of imagery and information products via web and media outlets
- ❑ Participate in MiTaM Working Group
- ❑ Participate in staff planning process
- ❑ Coordinate transition messages with US Embassy PAS and Affected State as appropriate
- ❑ Develop transition phase PAG and distribute to components, JTF, and US Embassy
- ❑ Track NGO and IGO operations for coordination and cooperation requirements
- ❑ Provide PA support to the commander
- ❑ Prepare the commander and other key staff members for media interviews
- ❑ Prepare news releases
- ❑ Refer media queries outside the scope of release authority to appropriate agencies or higher HQ
- ❑ Provide video and still imagery of military support to higher HQ in a timely manner

7.5.3.4 PAO Transition

- ❑ Maintain contact with US Embassy PAS
- ❑ Continue to facilitate media coverage
- ❑ Continue to coordinate with FDO on distribution of imagery and information products via web and media outlets

7.5.3.5 PAO Redeployment

- ❑ Implement a PA strategy for departure of military forces
- ❑ Develop a historical record of media coverage (video, photo, transcripts, etc.) gathered during the incident
- ❑ Prepare AAR comments and document lessons learned

7.5.4 Chaplain

Joint Publication (JP) 1, *Doctrine for the Armed Forces of the United States*, states that military commanders are responsible for providing the free exercise of religion to those under their authority.

JP 1-05, *Religious Affairs in Joint Operations*, defines the concept of "religious affairs" as consisting of two major capabilities of chaplains: religious advisement and religious support.

> **"** *Religious advisement is the practice of informing the commander on the impact of religion on joint operations to include, but not limited to: worship, rituals, customs, and practices of US military personnel, international forces, and the indigenous population.*
>
> *Religious support is Chaplain-facilitated, free exercise of religion through worship, religious and pastoral counseling services, ceremonial honors for the dead, crisis intervention, and advice to the commander on matters pertaining to morals, ethics, and morale as affected by religion."*
>
> *JP 1-05, Religious Affairs in Joint Operations*

Religious support in joint operations is dedicated to meeting needs of military and other authorized members in the personal free exercise of religion and providing commanders with professional advice regarding the dynamic influence of religion and religious belief in the operational area. The purpose of a Religious Support Team (RST) is to provide for, develop, and strengthen the spiritual and moral well-being of all members of the command.

Military chaplains, assisted by enlisted chaplain assistant support personnel, provide religious support as part of an RST, which normally consists of at least one chaplain and one enlisted support person. The RST deploys during FDR operations for the primary purpose of providing religious support to authorized US military personnel.

The Establishment Clause of the United States Constitution and current DOD legal guidance generally prohibits chaplains from providing religious support to the civilian population. However, following certain rare and catastrophic large-scale disasters, local and state capabilities of all types, to include spiritual care, may be overwhelmed. In these situations, an RST may serve as liaison to Non-Governmental Organizations (NGOs) and Faith-Based Organizations (FBOs) when directed by the commander.

7.5.4.1 Chaplain Preparation for Deployment

RST personnel should have training in Crisis Intervention Management tools such as Psychological First-Aid, Critical Incident Stress Management (CISM), Group Crisis Intervention and Disaster Behavioral Health, and should be thoroughly familiar with DOD Directive (DODD) 6490.5, "Combat and Operational Stress Control Programs," as well as Field Manual 4-02.51, Chapter 6, "Traumatic Event Management" and Air Force Instruction 44-153, *Traumatic Stress Response*.

Additional Resources for Managing Operational Stress

The Navy has the "Navy Leader's Guide for Managing Personnel in Distress."

The Marine Corps has the "Leaders Guide for Managing Marines in Distress."

Also refer to the webpage:
http://www.pdhealth.mil/op_stress.asp

❑ Establish RST to provide religious support to authorized DOD personnel
❑ Plan for operations and secure deployment of liturgical supplies
❑ During mass casualty events, identify coordination and planning requirements for chaplain activities with the Surgeon, the Red Cross and Red Crescent Movements, and other civilian agencies
❑ Coordinate with civilian ministry NGO/IGO organizations providing support
❑ Identify planning requirements between Service Component and civilian clergy
❑ Review AOR demographics to anticipate chaplain faith and denominational balance
❑ Plan for religious services
❑ Chaplains in supervisory positions will coordinate with appropriate staff agencies to ensure that subordinate chaplains and enlisted religious support personnel receive appropriate predeloyment training and education

7.5.4.2 Chaplain Deployment

- ❑ Service Components will identify, mobilize and/or deploy religious support personnel
- ❑ Maintain situational awareness of stress levels of assigned DOD personnel, first responders, and affected civilians, and take actions to provide care and mitigate stress
- ❑ Provide stress level situational awareness reports to the commander
- ❑ Be prepared to deploy with a minimum of 30 days of chaplain supplies

7.5.4.3 Chaplain FDR Operations

- ❑ Be prepared to conduct crisis intervention or Critical Incident Stress Management (CISM) training and services
- ❑ Conduct liaison and coordinate activities with other units
- ❑ Coordinate with NGOs and other agency religious personnel
- ❑ Provide religious support to authorized US Military and DOD civilian personnel and focus on mitigating the impact of traumatic events

7.5.4.4 Chaplain Transition

- ❑ RSTs advise the command on indicators, documenting civilian community capabilities to resume normal functioning without military support

7.5.4.5 Chaplain Redeployment

- ❑ Be prepared to conduct critical event debriefings or other CISM requirements
- ❑ RSTs conduct re-deployment religious support to assigned personnel and their families, focusing on reunion and reintegration issues with families
- ❑ Prepare AAR comments and document lessons learned

7.5.5 Knowledge Management

Knowledge Management (KM) integrates concepts from a variety of disciplines; information science, communication science, computer science and technology, social science, behavioral science, management science resulting in multiple definitions of KM. In this handbook, knowledge management is defined as *the art of creating, organizing, applying, and transferring knowledge to facilitate situational understanding and decision-making.*

In knowledge management, the distinction between data, information, and knowledge is significant. *Data* consists of pre-analyzed, objective facts that have no meaning in isolation. *Information* is data that has relevance and purpose; it is meant to inform or change the way a person understands something. Most simply, *knowledge* is the right information, at the right time, in the right context, available to the right person to support decision-making.

Of equal importance to knowledge management are the distinctions between Information Technology (IT), Information Management (IM), and Information Sharing. IT is the equipment/systems and storage devices used to support information services; IT is a KM enabler. IM is limited to organizing, managing, and distributing information. Information Sharing is about making information available. In short, KM is about linking people to people, information, and knowledge.

There are two different approaches to knowledge management: technology-centric and people/process-centric. Technology–centric approaches focus on explicit knowledge and do not address tacit knowledge. It centers on computers, where explicit knowledge or information is captured, stored, and accessed. People and process-centric approaches focus on tacit knowledge. It centers on the sharing of knowledge through person-to-person contact and interaction.

KM objectives:

- Provision information: with speed, agility, transparency, reliability and security
- Empower individuals: foster innovation, collaboration, and continuous improvement and provide knowledge, learning and cross-command situational awareness
- Enable Senior Leaders: provide them with timely knowledge to support decision making

KM goals:

- Ensure communication and coordination
- Encouraging elements to function as one mutually supporting command
- Understand where KM fits in the knowledge and information flow
- Use specific criteria for displaying common situational awareness
- Assess and filter external information
- Ensure rapid, easy and accurate retrieval of information and data
- Provide understandable products
- Avoid information overload and over-individualization
- Promote mutual trust and confidence

KM end state: a common operating picture, where network-centric data becomes visible, accessible, and understandable to all mission partners.

7-45

7.5.5.1 Knowledge Management Officer

In the military operations context, a Knowledge Management Officer (KMO) is a key horizontal and vertical staff facilitator. The KMO is responsible for developing and capturing the command's information management processes. This facilitates transforming information into knowledge and fosters appropriate access to and dissemination of commanders' information requirements. Doctrinally, the KMO answers to the JTF Chief of Staff.

7.5.6 Provost Marshal

The JTF Commander may determine a need to establish a JTF Provost Marshal (PM) position. The PM duties may include coordinating force protection functions with Affected State's law enforcement agencies and/or security personnel; conducting operational risk assessments; conducting terrorism vulnerability assessments; coordinating DOD law enforcement operations; establishing a customs pre-clearance program; or leading JTF force protection efforts.

7.5.7 Medical Officer/Surgeon/Medical Personnel

Medical personnel face unique challenges in a FDR environment. They must be prepared to provide support for both military and, if directed, civilians. Standard requirements include preventative medicine (PM), force health protection (FHP), and health service support (HSS) for US forces. Additionally, medical personnel may be directed to provide medical care for coalition partners or civilians and assist in stabilizing and reestablishing indigenous medical and public health resources affected by the disaster.

When a JTF is activated, a command surgeon (JTF-SG) is designated from one of the component Services. The JTF-SG assesses the HSS and FHP requirements and capabilities (including public health and medical needs) and provides guidance to the JTF Commander to enhance the effectiveness of medical support throughout the AOR.

Medical planners must be involved early in the planning process to ensure planned medical support to US forces is adequate and that planning of any directed FDR medical activities incorporates information from the Interagency and other potential partners. See JP 4-02 for additional information on medical planning process.

During FDR missions, medical forces may be tasked to assist local military and civilian health systems and provide direct health care support to include primary medical, dental, and veterinary care. To ensure mission success, medical personnel must partner with local medical authorities and the Affected State Ministry of Health to ensure that all support provided to the local populace is consistent with Affected State health care standards, applied in an equitable manner, and can be sustained by the Affected State.

Medical personnel must also maintain awareness of all other interagency, multinational, and local partners operating in the area to avoid duplication of effort, encourage synchronization of support, and ensure proper use of all available resources. Primary consideration must be given to the stabilization of existing medical infrastructure or establishing support when none exists. Commanders must ensure that no operations are undertaken that could have the effect of displacing existing medical infrastructure. Supporting or facilitating the restoration of extant medical infrastructure will help to build or restore confidence in the Affected State health care system and avoid disruption of services already being provided.

> **NOTE:** *There is some debate within the international community as to the usefulness of providing field hospitals since much of the need for medical support to the affected local population occurs prior to the arrival of US military forces. Using field hospitals may decrease collaboration with Affected State healthcare infrastructure and can potentially disrupt local medical practices and treatment protocols.*

Emergency medical treatment is a major concern within the first few hours following a disaster. Therefore, the Affected State's medical personnel will most likely provide support during the initial response. Proper assessments of the Affected State's health care system and infrastructure must be conducted prior to deployment to ensure that the right type and amount of medical assets are tasked to meet the needs of disaster victims. Proper assessments and coordination with the Affected State's Ministry of Health or other health care officials will also help to ensure compliance with local medical practices and treatment protocols and avoid disruption of services already being provided. For further guidance, the medical officer should review the WHO/PAHO Guidelines for the Use of Foreign Field Hospitals:

- "In the Aftermath of Sudden-Impact Disasters" retrieved at: http://new.paho.org/disasters/index.php?option=com_content&task=view&id=674&lang=en
- "Humanitarian Assistance in Disaster Situations: A Guide for Effective Aid" retrieved at: http://new.paho.org/disasters/dmdocuments/PED/Publications/books/pedhumen.pdf

- Editorial: Field Hospitals and Medical Teams in the Aftermath of Earthquakes retrieved at: http://new.paho.org/disasters/newsletter/index.php?option=com_content&view=article&id=429&Itemid=266&lang=en

7.5.7.1 Behavioral Health

Behavioral health personnel play a vital role in all US military operations. Behavioral health personnel advise leaders on preventive measures and provide assistance in addressing operational stress reactions and other behavioral health issues. During FDR operations, preventive measures must be implemented to ensure the health and wellbeing of US military personnel. In an FDR environment, it is important to maintain awareness of the stress levels of fellow Service members and subordinates. Commanders and leaders must ensure proper sleep (work and rest cycles), nutrition, exercise, and personal hygiene among Service members.

Handling human remains is a particularly stressful part of disaster relief efforts. While processing human remains is a mortuary affairs function, medical personnel (by virtue of their position) will likely encounter, be required to handle, or temporarily store the remains of deceased patients until mortuary affairs personnel can be contacted.

The following are examples of coping strategies that can be used to minimize operational stress reactions when handling human remains:
- Remembering the greater purpose of the work
- Talking with others and listening well
- Using humor to relieve stress (avoid personal or inappropriate comments)
- Not focusing on individual victims
- Getting teams together for mutual support and encouragement
- Providing opportunities for voluntary, formal debriefings

7.5.7.2 General Health Risks

After a disaster, the civilian population and military personnel have an increased risk of exposure to illnesses spread by contaminated food or drinking water; vector-borne diseases (i.e., those diseases spread by mosquitoes, lice, rodents, etc.); close contact with sick or injured individuals; high levels of industrial pollution; stress; fatigue; and regionally endemic diseases (e.g., malaria, dengue, Japanese Encephalitis, etc.). The JTF must have robust preventive medicine assets to conduct medical and environmental health risk assessments and identify effective preventive medicine measures to counter the threat to US forces. Health risks rise with:
- Lack of proper refuse and human waste disposal

7-48

- Contaminated food
- Contaminated water
- Inadequate water for hygiene
- Increased exposure to the heat or cold

7.5.7.3 Medical Officer Pre-Deployment

❑ Be prepared to provide a Health Care Provider (HCP) and/or preventative medicine specialists for deployment with the advance party

❑ Conduct Medical Intelligence Preparation of the Operational Environment (MIPOE) with J2; helpful resources include but are not limited to:

  o Embassy Emergency Action Plan (especially helpful for facility assessments)
  o United States Army Public Health Command
  o National Center for Medical Intelligence
  o Centers for Disease Control and Prevention
  o World Health Organization and others

❑ Locate and obtain pertinent information on medical capabilities in AOR

❑ In coordination with USAID/OFDA, J-2, and J-3, determine NGO medical capabilities and needs

❑ Provide interpretation of current medical capabilities and status of Affected State health care infrastructure

❑ Conduct assessment of health threats of operational significance, assess available medical support resources, and plan for mitigation of health threats prior to deployment

❑ Establish evacuation policy for patient movement within the JOA

❑ Develop patient tracking system

❑ Prepare mass casualty plan

❑ Prepare medical annex of the OPORD

❑ Estimate medical logistics requirements and preplan Class VIII resupply sets or preconfigured push-packages to support initial sustainment operations until replenishment by line-item requisitioning is established

❑ Ensure that medical personnel are trained and knowledgeable of health threats and diseases prevalent in the region as well as language, social, religious, cultural, and political factors that may impact the provision of medical support

❑ Plan and conduct medical portion of personnel readiness processing, to include all necessary vaccinations and prophylactic medications

7-49

- ❑ Ensure compliance with all OSD/HA Force Health Protection and Readiness requirements and theater specific deployment health guidance
- ❑ Ensure personnel deploying longer than 30 days and reserve component personnel activated for longer than 30 days must complete DD Form 2795. Only those reservists activated for less than 31 days are exempt from completing the form
- ❑ Ensure that all military personnel deploy with a 90-day supply of individual prescription medication
- ❑ Prepare for medical portion of Joint Reception, Staging, and Onward movement, and Integration (JRSOI)
- ❑ Recommend task organization of medical elements to satisfy mission requirements
- ❑ Plan and implement medical support operations to ensure the provision of appropriate roles of care

NOTE: JTF commanders typically designate the Army component to serve as the Single Integrated Medical Logistics Manager (SIMLM). The Theater Lead Agent for Medical Materiel (TLAMM) and SIMLM work together to develop the medical logistics support plan for synchronization of medical requirements and Class VIII distribution to the JTF. Refer to JP 4-02 and FM 4-02.1 for additional information.

7.5.7.4 Medical Officer Deployment

NOTE: Ensure that medical stability operations are conducted in accordance with DOD Instruction (DODI) 6000.16.

- ❑ Identify the medical and public health-related Commander's Critical Information Requirements (CCIR) and provide updates
- ❑ Coordinate with the US Embassy Country Team or Affected State Ministry of Health and Ministry of Agriculture to gain access to assessments or information that may be available regarding the state of Affected State health care and agricultural resources and infrastructure
- ❑ If directed, establish Medical Treatment Facility (MTF) or aid stations within the AOR
- ❑ Monitor status of medical and public health support to joint, interagency, intergovernmental, and multinational partners

❑ Determine medical workload requirements (patient estimate) based on J-1 estimate

❑ Ensure that DSCA approval is obtained prior to transferring or issuing US medical supplies and equipment to the Affected State

❑ Coordinate with the chaplain or RST for information regarding religious based dietary restrictions for hospitalized patients

❑ Monitor the assignment, reassignment, and use of medical personnel within the AOR

❑ Plan and implement medical support operations to ensure the provision of appropriate roles of care

❑ Prepare MEDSITREP daily as directed by higher HQ (format can be found in Appendix C)

NOTE: *A complete Joint Operational Planning and Execution System(JOPES) MEDSITREP can be found at the NORTHCOM Surgeon portal:*
https://operations.noradnorthcom.mil/sites/CommandSpecialSt aff/SG/MedPlansOpsDiv/WorkSpace/Shared%20Documents/ MEDSITREP%20Final%206Apr07.doc

❑ Medical and safety personnel recommend site and mission specific personal protective equipment (PPE)

❑ Issue policies, protocols, and procedures pertaining to eligibility for care (medical, dental and veterinary treatment) for sick or injured

❑ Conduct medical surveillance to assess health threats of operational significance, assess available medical resources, and plan for the mitigation of health threats

❑ Ensure that appropriate preventive medicine and environmental health capabilities are employed to support casualty prevention and protection of the force from health threats

❑ If the decision is made to move casualties out of the Affected State, coordinate with Joint Patient Movement Requirements Center for patient movement

❑ Ensure the documentation of medical encounters and health hazard exposures as part of the patient's individual health record (either electronically or on paper medical records)

❑ Coordinate for reach-back support and staff augmentation as required

❑ In coordination with Affected State MTFs, determine transition plan for disposition and filing of civilian medical records

- ❑ As directed, provide veterinary personnel to assist in the evacuation, triage, and medical treatment of military working dogs and other government owned animals
- ❑ Coordinate with the Armed Services Blood Program Office to provide available blood products

7.5.7.5 Medical Officer FDR Operations

- ❑ Prepare reporting and regulating instructions in support of FDR operations
- ❑ Update and disseminate information regarding emerging medical threats
- ❑ Provide medical evacuation support (air and ground) of seriously ill or injured patients
- ❑ Provide medical surveillance and laboratory diagnostic testing
- ❑ Provide preventive medicine support for the inspection of US military, multinational, and when directed, Affected State water sources and food storage facilities
- ❑ When directed, provide technical assistance to establish public health and sanitation, and training and education programs for the local populace
- ❑ Ensure that the proper approval is obtained prior to the transfer or issue of US medical supplies and equipment to the Affected State
- ❑ Provide available medical teams for casualty clearing/staging and other medical support missions as directed
- ❑ Provide available logistical support to medical and public health response operations
- ❑ Participate in MiTaM Working Group
- ❑ Provide preventive medicine personnel to assist in activities for the protection of public health (such as food, water, waste water and solid waste disposal, vectors, hygiene, medical waste disposal, incineration, and other environmental conditions)
- ❑ Coordinate with Affected State mortuary affairs personnel for disposition of human remains

7.5.7.6 Medical Officer Transition

- ❑ Conduct ongoing assessments of support
- ❑ Ensure effective transfer of patients, and patient records to Affected State medical authorities
- ❑ Assist in planning follow-on support or transition responsibility to the Affected State government, joint, interagency, intergovernmental, or multinational partners

7.5.7.7 Medical Officer Redeployment

- ❑ Ensure adequate transfer of logistics support, evacuation information, and personnel support prior to transfer of responsibility to civilian MTF and redeployment
- ❑ Provide morbidity and mortality data to Affected State medical authorities as part of the transition
- ❑ Assist with Line of Duty (LOD) investigations
- ❑ Plan and conduct Post-Deployment Health Risk Assessment (PDHRA)
- ❑ Prepare AAR comments and document lessons learned

This page intentionally left blank

CHAPTER 8 - SAFETY

8.1 Overview

Risk management, safety, accident prevention, and force health protection are critical to mission accomplishment. Leaders at all levels, Service personnel, and civilians are all responsible for safety.

8.2 Risk Management

Commanders are responsible for assessing their operation as a total system. During mission planning and execution, leaders should proactively identify hazards, assess associated risks, and develop control measures necessary to reduce risk.

The US Department of Defense summarizes the principles of Operational Risk Management (ORM) as follows:

- Accept risk when benefits outweigh the cost
- Accept no unnecessary risk
- Anticipate and manage risk by planning
- Make risk decisions at the right level

Commanders and their staffs should constantly assess the operational environment and ask questions such as the following:

- What are the human, mechanical, and environmental hazards associated with this operation?
- Can hazards be eliminated or at least mitigated?
- What are the risks associated with hazards if they cannot be mitigated or eliminated?
- Are there safer options that may be pursued?
- Do the potential benefits outweigh the risks?

Hazard identification, mitigation, and management of risk are key factors in safely conducting FDR operations. The Army uses Composite Risk Management (CRM), while the other Services use ORM. Both programs identify hazards associated with the conduct of operations, assist in mitigation of associated risks, and management of residual risks.

Department of Defense Instruction (DODI) 6055 (series) is the basis for DOD Safety and Occupational Health Programs.

The Army Safety Program is addressed in Army Regulation (AR) 385-10; Department of Army Pamphlet (DA PAM) 385-1; Army Field Manual (FM) 5-19; and the Composite Risk Management FM 5-0.

The United States Air Force safety program is detailed in Air Force Instruction (AFI) 91-301 and AFI 90-901, Operational Risk Management.

The United States Navy and Marine Corps use the Office of the Chief of Naval Operations Instruction (OPNAVINST) 5100 series, OPNAVINST 3500.39B, and the Marine Corps Institute (MCI) Operational Risk Management Manual (ORM 1-0).

Service's Safety Websites:

US Army Combat Readiness Safety Center
https://safety.army.mil/

US Navy Safety Center
http://www.safetycenter.navy.mil/index.asp

US Air Force Safety Center
http://www.afsc.af.mil/

8.2.1 Composite Risk Management

CRM is a decision-making process used to mitigate risks associated with all hazards that have the potential to injure or kill personnel, damage or destroy equipment and resources, or otherwise negatively impact mission effectiveness. The underlying assumption of CRM is that risks can be quantified and managed by determining the probability and severity of a potential loss due to any number of hazardous conditions or situations that may be encountered in the operational environment.

CRM is a 5-step process:
1. Identify hazards to the force
2. Assess hazards to determine risks
3. Develop controls and make risk decisions
4. Implement controls
5. Supervise and evaluate

8.2.1.1 Definitions of Severity

- *Catastrophic*: death or permanent disability, system loss, major property damage
- *Critical*: permanent partial disability, temporary total disability in excess of three months, major system damage, significant property damage
- *Moderate*: minor injury lost workday accident, compensable injury or illness, minor system damage, minor property damage
- *Negligible:* first aid or minor supportive medical treatment, minor system impairment

8.2.1.2 Definitions of Probability

- *Frequent*: occurs often, continuously experienced
- *Likely*: occurs several times
- *Occasional*: occurs sporadically
- *Seldom*: unlikely, but could occur at some time
- *Unlikely*: can assume it will not occur

8.2.1.3 Definition of Risk Levels

Risk levels are a combination of severity and probability of hazards (see Figure 8-1).

- *Extremely High (E)*: loss of ability to accomplish mission if hazards occur during mission
- *High (H)*: significant degradation of mission capabilities if hazards occur during mission
- *Moderate (M)*: minor degradation of mission capabilities; reversible
- *Low (L)*: little or no impact on mission accomplishment

8.2.1.4 Risk Assessment Matrix

Figure 8-1 provides leaders with a process to assess risk based upon severity and probability of hazards. The point at which the severity row and the probability column intersect defines the level of a hazard. Taking unnecessary risks is unacceptable.

Probability / Severity		Frequent A	Likey B	Occasional C	Seldom D	Unlikely E
Catastrophic	I	E	E	H	H	M
Critical	II	E	H	H	M	L
Moderate	III	H	M	M	L	L
Negligible	IV	M	L	L	L	L

Figure 8-1: Risk Assessment Matrix

8.2.1.5 Typical Army Risk Acceptance Levels:

- *Low Risk*: typically approved at the Company (tactical) level and below
- *Moderate Risk*: typically approved at the Battalion (tactical) level
- *High Risk*: should be approved at the Brigade (operational) level
- *Extremely High Risk*: must be approved at the Geographic Combatant Commander (GCC) (strategic) level although may be delegated to the JTF Commander (operational) level

8.2.2 Operational Risk Management

ORM is cyclical process, which includes risk assessment, risk decision-making, and implementation of risk controls, which results in acceptance, mitigation, or risk avoidance. ORM is a decision making tool used by personnel at all levels to increase effectiveness by identifying, assessing, and managing risks. By reducing the potential for loss, the probability of a successful mission is increased.

8.2.2.1 Principles of ORM

There are four basic principles that provide the foundation for ORM and the framework for implementing the process.

- Accept risk when benefits outweigh the cost
- Accept no unnecessary risk
- Anticipate and manage risk by planning
- Make risk decisions at the right level

8.2.2.2 Three Levels of ORM

- *In-depth* risk management is used before a project is implemented, when there is plenty of time to plan and prepare. Examples of in-depth methods include training, drafting instructions and requirements, and acquiring personal protective equipment.

- *Deliberate* risk management is used at routine periods through the implementation of a project or process. Examples include quality assurance, on-the-job training, safety briefs, performance reviews, and safety checks.
- *Time-critical risk management* is used during operational exercises or execution of tasks and is defined as the effective use of all available resources by individuals, crews, and teams to safely and effectively accomplish the mission when time and resources are limited.

8.3 Safety

The safety program applies to all DOD personnel and is comprised of several elements, including force health protection, accident prevention, and reporting.

8.3.1 Force Health Protection

Joint Publication 1-02 defines *force health protection* (FHP) as "Measures to promote, improve, or conserve the mental and physical well-being of Service members".

Addressing FHP measures for the FDR environment is critical to mission success. Commanders should:

- Obtain health threat information from the National Center for Medical Intelligence (NCMI) regarding specific FHP measures for the geographic area
- Direct implementation of preventive medicine measures (such as vaccinations and other chemoprophylaxis) during pre-deployment for forces entering at-risk locations to counter health threats
- Ensure deploying personnel have access to potable water and are only provided with nutritional resources procured from approved sources
- Ensure deploying forces have access to proper personal protective equipment (e.g., insect repellents with DEET, permethrin spray, mosquito netting, eye and hearing protection)
- Take necessary precautions for personnel involved in physical labor to prevent sun exposure and heat-related injuries or excessive exposure in cold environments
- Ensure proper hygiene and sanitation measures as well as trash, medical, and human waste disposal procedures are followed
- Consider deployment of mental health teams in disasters involving mass fatalities

> **NOTE:** *Military personnel who have not been exposed to disaster environments may develop post-traumatic stress disorder-related issues. Early intervention and counseling can help to mitigate those issues.*

- Request medical teams specifically monitor response personnel for signs of illness and stress
- Consider rotating personnel away from the impacted area occasionally if they are in the impacted areas for extended periods

8.3.2 Fatigue

The work pace in response to disasters is demanding and leaders should ensure personnel avoid physical exhaustion. Rotating personnel between demanding and less demanding tasks can mitigate fatigue.

8.3.3 Personal Hygiene

Personal hygiene requires special attention. Many natural and man-made contaminants pose potential risks to personnel during FDR operations. Precautions to minimize those risks include ensuring the availability of potable water, and when practicable, the use of functional latrines, shower and washing stations, laundry, and waste disposal facilities. Personnel should be encouraged to wash their hands frequently to reduce the risk of disease transmission.

8.3.4 Food Safety

Food safety is paramount for maintaining a healthy force in the face of a disaster. Disasters are extremely disruptive and are likely to produce environments in which contamination can occur with exceptional ease. Prolonged exposure to moderate levels of heat can quickly lead to food spoilage and cause widespread acute gastrointestinal distress in field-deployed forces. Foodstuffs that are exposed to chemicals, radiation, biohazards, pest infestations, toxic industrial materials, smoke, and flooding that has been released into the environment as a result of disasters also pose significant risks to the health of the force.

8.3.5 Preventable Injuries

Appropriate safety gear can prevent many injuries to eyes, ears, heads, hands, back, and feet of FDR personnel. When appropriate, personnel should wear protective lenses, goggles, face shields, hearing protection when operating, or exposed to, heavy equipment and machinery. Helmets or hard hats should be worn in debris-filled areas.

Operational personnel should not wear jewelry to avoid injuries such as electrical shock, de-gloving injuries, or amputations. Personnel should wear gloves, use proper lifting techniques and equipment to avoid back injuries. Additionally, personnel should wear footwear suitable for hazardous environments and follow preventative measures for trench foot and fungal infections.

8.3.6 Respiratory Hazards

Respiratory hazards (smoke, ash, molds, various airborne contaminants and toxic chemicals) are common to disasters. Personnel may be exposed to asbestos, carbon monoxide, nuisance dust, and caustic vapors and should use appropriate military or civilian gas, mist, fume, or dust protective masks. FHP personnel should conduct tests to identify hazards and widely disseminate information to the force.

> **WARNING:** *Improper removal of debris containing mold, asbestos, and radiation can have serious impacts upon the health of Service members.*

8.3.7 Blood-borne Pathogens and Disease

FDR personnel may be at risk to potential endemic pathogens, to which they may not have immunity. Additionally, they may be exposed to airborne, contact, and blood-borne pathogens secondary to the rescue, movement, and treatment of disaster victims. Personnel should have up-to-date immunizations, based upon available medical intelligence regarding endemic diseases, access to appropriate chemo-prophylactic medicines, and should observe basic preventive medicine precautions to minimize disease risk.

> **NOTE:** *Personnel should use the following equipment whenever required:*
> - *Latex or rubber gloves*
> - *Over-garments for added protection*
> - *Face masks for respiratory protection*
> - *Goggles for eye protection*
> - *Biohazard bags*
>
> *(Biohazard bags/regulated medical waste bags should only be used by personnel qualified in handling hazardous materials)*

8.3.8 Stress

Military personnel involved in FDR operations may experience stress and anxiety. Medical Combat and Operational Stress Control (COSC) teams, and chaplains, are trained to assist personnel with stress management. Stress management support channels for civilians may include local religious institutions and the Red Cross or Red Crescent Societies. Selected NGOs may have the ability to send stress management teams to help citizens affected by the disaster.

> *For information on the control of stressors and actions to control stress, see FM 4-02.51, Combat and Operational Stress Control and FM 6-22.5, Combat and Operational Stress Control Manual for Leaders and Soldiers.*

8.3.9 Animal Hazards

Disaster conditions may increase the risk of FDR personnel to bites and scratches from animals, including rats and venomous snakes, which may increase the danger from diseases such as rabies. Additionally, personnel may become infested with lice and fleas. Household pets may become more aggressive or dangerous than usual. Accordingly, personnel should take precautions to avoid animal and snakebites, and not taunt, play with, or handle any animals.

8.3.10 Biting or Stinging Insects and Spiders

Personnel should be aware of the potential risk presented by mosquitoes, ticks, chiggers, ants, venomous spiders, fleas, lice, wasps, and bees. Personnel should make use of mosquito netting, treat clothing with permethrin, and use appropriate countermeasures as directed, and supervised, by medical authorities.

> *For more information on animal and insect-borne diseases, refer to the US Army Public Health Command website at http://phc.amedd.army.mil/topics/envirohealth/epm/Pages/default.aspx or http://phc.amedd.army.mil/Pages/default.aspx*

8.3.11 Hazardous Plants

Personnel may encounter hazardous plants in FDR environments, some of which may be poisonous. They may require special handling and observance of safety procedures. Some species of plants, such as oleander

8-8

and poison ivy, are poisonous to the touch and burning them will release
toxic chemicals. Refer to the Army Center for Health Promotion and
Preventive Medicine website at
http://phc.amedd.army.mil/Pages/default.aspx

8.3.12 Use of Vehicles and Transportation

Personnel should take extra care to drive defensively in disaster-impacted
areas and remain alert to the potential for significantly increased hazards in
areas that would have otherwise been considered safe for driving.
Operators of vehicles should adhere to the following rules:

- The senior occupant is responsible for all vehicle safety
- Pair experienced drivers with inexperienced drivers for supervision
  and hands-on training
- Ensure proper use of seatbelts
- Use experienced drivers in difficult terrain
- Remind drivers to slow down in limited visibility, on rough terrain,
  and during inclement weather
- Secure vehicle antennas to prevent contact with power lines and
  other objects
- Take into account the maximum fording depth for each vehicle
  type, and ensure proper fording equipment and accessories are
  installed before entering water areas
- Use ground guides during periods of limited visibility
- Ensure operators are licensed on their vehicle; operators designated
  to transport Hazardous Materials (HAZMAT) are licensed to load,
  transport and off-load those materials

All operators of vehicles should perform:

- Preventive maintenance checks and services, especially under
  adverse or unusual conditions
- Special requirements covered in the "Operating Under Unusual
  Conditions" section of their respective operator's manual

Leaders should conduct convoy briefings before movement. Leaders
should ensure all vehicle operators know how to:

- Conduct a physical reconnaissance of the route to avoid hazards;
  mark unavoidable hazards on a strip map and include them in the
  convoy briefing
- Reconnoiter the route for bridges or underpasses that might be too
  low for large vehicles
- Access roads, bridges, and overpasses that may not be posted with
  weight or height restrictions

- Reconnoiter routes for hazards below the water line before operations begin
- Check water height before driving on submerged surfaces; a good rule of thumb is not to drive into running water deeper than the vehicle axle

Convoys require a safety briefing containing, at a minimum, the following:

- Mishap duties and responsibilities
- Speed limits
- Interval distances
- Mechanical breakdown procedures
- Passenger safety measures
- Visual signals for convoy halt, caution, slow, etc.
- Preplanned rest halts
- Hospital and operational mission support locations identified on a provided strip map as applicable

Drivers will not operate a vehicle for longer than two hours without a rest stop or four hours without relief.

8.3.13 Accident Reporting

All accidents will be reported within 24 hours to the task force safety office. At a minimum the following information is provided for each accident reported:

- Name of the person reporting the accident
- Point of contact telephone number
- Unit involved in the accident
- Location of the accident
- Date and time of the accident
- Name and rank of personnel involved
- Extent of injuries
- Type of property or equipment damage
- Estimated cost of damage
- Estimated environmental cost

8.4 Areas of Special Concern

The following are areas of special concern that may require additional/special planning (not all-inclusive):

- Night operations
- Aircraft operations
- Water operations
- Weapons (if necessary, both military use and civilian use)

- Tactical rest policy
- Field heaters and stoves (if applicable)
- Petroleum, Oil, and Lubricants (POL) storage and handling
- Hazardous Materials
- Unexploded munitions (if applicable)
- Hot weather/cold weather operations

8.5 Hazards Identification and Mitigation

To identify hazards, leaders should obtain information about the characteristics of the specific geographical region and overall effects of the disaster. For example, flooding of buildings has significant secondary effects in hot, humid environments. Toxic mold and fungus thrive in these conditions. Standing, water-damaged structures can become uninhabitable for humans but may shelter dangerous stray or wild animals, insects, and reptiles.

8.5.1 Electrical Hazards

All electrical transformers pose severe risks. Electrical lines can present a lethal shock hazard. To avoid injuries:

- Do not attempt to move transformers during cleanup
- Mark transformers and report locations to the chain-of-command
- Do not touch, work, or operate equipment near downed power lines; electricity might be restored to downed power lines without notice
- As commercial power is re-supplied, all emergency generators should be taken offline. Only qualified utility or engineer personnel conduct the changeover. If a downed power line is difficult to see but is in a traffic area, clearly mark the area so no one touches the downed wire. Use caution when antennas are near power lines and avoid erecting antennas near power lines. Identify antennas that may fall on power lines or on people and take appropriate action to prevent accidents or injury.

WARNING: *Always assume downed power lines are live. Water (including snow) is an excellent conductor of electricity. Stay away from downed power lines of any kind to avoid electrical shock or grave injury.*

8.5.2 Power Generator Safety

Generator usage during operations can create special concerns. Personnel entering homes and buildings need to be aware of the potential carbon monoxide threat posed by generators used indoors that do not properly vent

8-11

exhaust outside of an enclosed area. Military personnel using generators must give special attention to the following:

- Operation only by trained personnel
- Fueling operation hazards
- Proper grounding and bonding of generators
- Carbon monoxide hazards
- Generator fire hazards and fire protection
- Generator electrical load limits and capacity
- Electrocution hazards, prevention, and first aid

> **WARNING:** *Military personnel are not permitted to connect military generators to civilian infrastructure!*
>
> *A certified electrician must be available to connect the power. Ensure that power lines are not re-energized by connecting infrastructure to generators.*

8.5.3 Handling Contaminated Items

Disasters can cause massive infrastructure disruptions that can facilitate the release of Toxic Industrial Chemicals and Materials (TICs/TIMs), fuels, sewage, and potentially radioactive material into the environment. Take precautions when handling and collecting materials that have potentially been contaminated by these types of dangerous exposures. A collection site for contaminated items should be established. In addition, sites should be designated for showering and clothing changes before moving personnel to non-contaminated areas. For more information, see the following websites:

- US Department of Labor Occupational Safety and Health Administration: www.osha.gov
- DOD Chemical, Biological, Radiological and Nuclear Defense Information Analysis Center: http://www.CBRNEiac.apgea.army.mil
- Center for Disease Control and Prevention: http://www.cdc.gov
- US Army Maneuver Support Center: www.wood.army.mil

8.5.4 Fire

Units should deploy with sufficient fire extinguishers and fire prevention equipment for organic operations and living areas. For further information, refer to Fire Rescue I at http://www.firerescue1.com.

CHAPTER 9 - REGIONAL RESPONSE ORGANIZATIONS

9.1 Overview

This chapter provides an overview of Geographic Combatant Commands (GCC). This chapter is presented in two parts: the first part identifies the common elements of FDR response at the GCC level; the second part presents unique elements of the GCC by region.

9.2 Part One: Unified Combatant Commands (Geographic)

JP 1-02 defines a *unified combatant command* as a command with a broad continuing mission under a single commander and composed of significant assigned components of two or more Military Departments, established and so designated by the President, through the Secretary of Defense with the advice and assistance of the Chairman of the Joint Chiefs of Staff.

There are two types of Unified Combatant Commands (UCC): geographic and functional. This chapter describes the geographic combatant commands. The Functional UCCs are described in Chapter 10. Commanders of unified combatant commands are four-star flag or general officers, who are assigned a geographic area of responsibility.

The GCC are:

- US Africa Command (USAFRICOM)
- US Central Command (USCENTCOM)
- US European Command (USEUCOM)
- US Northern Command (USNORTHCOM)
- US Pacific Command (USPACOM)
- US Southern Command (USSOUTHCOM)

GCCs can function in either of two capacities, as a *supported commander* or as a *supporting commander.*

A *supported commander* has primary responsibility for all aspects of assigned missions. That commander may receive assistance from another commander's force and is responsible for ensuring that supporting commanders understand the assistance required.

A *supporting commander* may provide augmentation forces or assets to a supported commander and is responsible for providing the assistance requested.

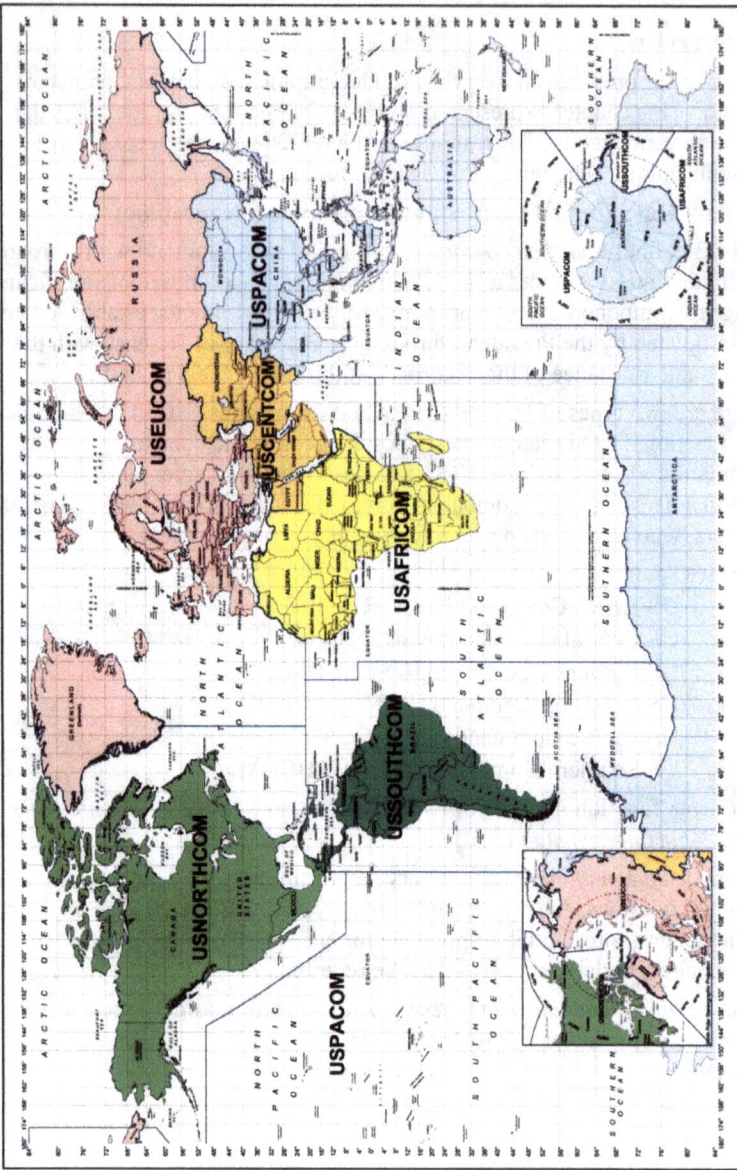

Figure 9-1: GCC Areas of Responsibility

UNCLASSIFIED

In response to a disaster, the supported GCC determines the forces necessary to conduct and sustain the FDR operation, which may involve forming a Joint Task Force (JTF).

Each GCC has a Crisis Action Team (CAT) or rapid deployment team that is initially deployed as the immediate responder/assessor for the GCC. In addition, a GCC may choose to utilize various organization entities or resources such as the Humanitarian Assessment Survey Team (HAST), a Civil-Military Operations Center (CMOC), Humanitarian Assistance Coordination Center (HACC), and Liaison Officers (LNOs) to coordinate and facilitate the FDR response. The exact composition of those teams and the subsequent follow-on assets will vary depending on the type and severity of the incident and, in some cases, restrictions emplaced by the Ambassador or by the Affected State. A brief description of the assets and resources available to support the JTF are described in the following sections.

9.2.1 Crisis Action Team

During the assessment process, special teams are assembled at the level where the problem is being assessed and its resolution is being developed. A Crisis Action Team (CAT) includes representatives from all command and staff divisions, and may include representatives from a wide range of involved organizations. It is activated by the GCC J-3 and handles matters that exceed the capability of an initial response cell. The CATs assigned functions include:

- Generating, exchanging, and receiving information
- Developing military options, courses of action, and concepts of operation
- Orchestrating and monitoring deployment of response forces

9.2.2 Humanitarian Assistance Survey Team

The supported GCC may organize and deploy a Humanitarian Assistance Survey Team (HAST) to acquire information required for planning. Once deployed, the HAST, working with the country team and USAID, can assess the ability of the Affected State government to respond to the disaster; identify primary points of contact for coordination and collaboration; determine the threat environment for force protection purposes; survey facilities and infrastructure that may be used to support FDR operations including Aerial Port of Debarkation (APOD), Seaport of Debarkation (SPOD), Forward Operating Base (FOB); conduct Health Service Support (HSS) and Force Health Protection (FHP) assessments and coordinate specific support arrangements for the delivery of food and medical supplies. Suggested team composition should include:

- Medical personnel qualified to conduct HSS and FHP assessments (including environmental vulnerability assessments)
- Engineers (structural, mechanical, etc.)
- Logisticians
- Communications experts
- Transportation management specialists
- Force protection experts

9.2.3 Joint Inter-Agency Coordination Group

The Joint Inter-Agency Coordination Group's (JIACG) primary role is to enhance interagency coordination, and to collaborate at the operational level with other USG civilian agencies and departments. It may play an important role in contingency planning for FDR and in initial interagency coordination prior to establishment of a HACC or other coordination body.

9.2.4 Joint Logistics Operations Center

The Joint Logistics Operations Center (JLOC) receives reports from supporting commands, Service components, and external sources and collates information for decisions and briefings. Additionally, the JLOC coordinates all logistics, maintenance, transportation, and supply activities.

9.2.5 Joint Deployment Distribution Operations Center

The Joint Deployment Distribution Operations Center (JDDOC) links strategic deployment and distribution processes to operational and tactical functions.

9.2.6 Joint Task Force

A JTF Commander is normally assigned a joint operations area (JOA) in the combatant commander's area of responsibility (AOR). The JTF Commander and staff operate primarily at the operational level. The following is a partial list of working groups that may be formed by the JTF:

- Joint Collection Working Group
- Joint Logistics Working Group
- Rules of Engagement Working Group
- MiTaM Working Group
- Effects Assessment Working Group
- Humanitarian Assistance Working Group

9.2.7 Civil-Military Operations Center

The Civil-Military Operations Center (CMOC) is an organic component of Civil Affairs units. CMOCs provide operational level coordination between the JTF Commander and other stakeholders. It is established by the J-9 and run by Civil Affairs personnel (Army, Navy, or USMC) organic to the JTF.

The responsibilities of the CMOC need to be established quickly during joint operational planning. While sharing many general characteristics, each FDR operation is unique, and the CMOC structure must be tailored for each emergency.

In support of USAID/DART and the Affected State, the CMOC staff assists in coordinating military efforts and resources with the international humanitarian community. The JTF may include support from the following liaisons: other governmental agencies, US Army Corps of Engineers, key NGOs, IGOs, and the Affected State. See Appendix B for more information on DOD interaction with NGOs.

9.2.8 Joint Forces Air Component Commander

The JFACC is responsible for US Military air operations, and the coordination of those operations with the US Embassy, USAID/OFDA, the Affected State, the UN, and the International Civil Aviation Organization (ICAO). The JFACC will issue guidance describing processes and procedures for safe employment of US Military air assets operating within the disaster area. The JFACC may establish one or more Air Component Coordination Element (ACCE) teams to integrate air operations within the joint task force. See Section 7.4.6 for a description of the ACCE.

9.2.8.1 Airspace Command and Control

Aviation operations following a natural disaster can be complex and challenging. The effective management of air operations can be a key facilitator to the success of the overall FDR operation.

The Affected State maintains control of their airspace. They have final authority on matters of air traffic safety and management, including: establishment and management of temporary flight restrictions; development and implementation of aviation incident response plans; and coordination with air traffic control facilities. If the Affected State has lost the ability to manage their airspace, they may request airspace management support from either the UN World Food Program or the JTF ACCE.

An aviation and airspace management committee may need to be established to provide unified planning that integrates all aviation resources of agencies participating in response efforts. The aviation and airspace committee should be linked to the Affected State's aviation authority. This committee should provide guidance on:

- Management of airspace, airfields, helipads, and landing strips
- Prioritization of aviation missions
- Support of air mission requests
- Mission tasking of available aircraft
- Air mission planning and coordination including deconfliction

- Situational awareness of aviation operations in the disaster area

Coordination of ground support at designated airfields and airports

9.2.8.2 Daily Air Tasking Order

Military aircraft in the Joint Operations Area (JOA) are tasked via an Air Tasking Order (ATO), which includes applicable mission information in the Airspace Control Plan and Special Instructions (SPINS) section of the ATO. All participating military aircraft are required to adhere to the Airspace Control Plan and applicable ATO SPINS.

9.2.8.3 Regional Air Movement Coordination Center

The Regional Air Movement Coordination Center (RAMCC) coordinates with USAID/OFDA and the Affected State to schedule and manage DOD air assets within the JOA. The RAMCC coordinates with military command and control elements, such as Contingency Response Group/Element/Team (CRG/CRE/CRT), present at airfields to maximize logistics throughput at designated Aerial Ports of Debarkation (APOD). The RAMCC then coordinates with appropriate airspace control authorities to determine landing and takeoff times for US military aircraft.

9.2.9 Joint Personnel Recovery Center

Currently, the Commander of US Joint Forces Command (USJFCOM) is DOD's executive agent for Personnel Recovery (PR). See chapter 10 for a discussion on the dissolution of USJFCOM. The Joint Personnel Recovery Agency (JPRA), headquartered at Fort Belvoir, Virginia, is a subordinate activity of US Joint Forces Command's (USJFCOM) Joint Training Directorate/Joint Warfighting Center. USJFCOM has designated JPRA as the office of primary responsibility for DOD-wide personnel recovery matters. In that role, the agency supports the military departments, combatant commands (COCOM), the Joint Staff, the Office of the Secretary of Defense, its agencies, DOD field activities and other government agencies. JPRA enables commanders, individuals, recovery forces, and supporting organizations to accomplish five PR tasks: report, locate, support, recover, and reintegrate. The agency also assists COCOMs in developing and exercising PR architectures and operational procedures to plan and prepare for the recovery of isolated personnel.

9.2.9.1 Personnel Recovery

Individuals can become inadvertently separated from their organizations. DOD mandates recovery of isolated personnel when it involves DOD personnel (military, civilians, or contractors). The Joint Personnel Recovery Center (JPRC) assists in coordinating recovery efforts of US personnel who have become isolated. The JTF should consider all the capabilities available when planning and executing PR missions, including Affected State security organizations, NGOs/IGOs, as well as the resources within the local population. Recovery of isolated personnel during FDR is subject to established authorities, Status of Forces Agreements (SOFA), and Rules of Engagement.

> **NOTE**: *The Chief of Mission may request DOD assistance with the recovery of other civilians, such as international aid staff, but that is usually a matter for local authorities.*

9.2.10 Geospatial Planning Cell

USCENTCOM, USEUCOM, USPACOM, and USSOUTHCOM each have a Geospatial Planning Cell (GPC). Neither USAFRICOM nor USNORTHCOM have a supporting GPC. A GPC is responsible for geospatial operational planning. The detachment is responsible for generating and analyzing terrain data, preparing decision graphics, image maps, and 3D terrain perspective views. A key task for the GPC is the management of the Theater Geospatial Database, map updates, tactical decision aids, and other decision support graphics. It is capable of coordination with external geospatial assets such as National Geospatial Intelligence Agency (NGA) and the Army Geospatial Center, and coordination with Affected State topographic support activities, and higher headquarters.

9.2.11 Humanitarian Assistance Coordination Center

The Humanitarian Assistance Coordination Center (HACC) is a temporary body that operates during the early planning and coordination stages of an operation. See Appendix B for more information on the HACC.

9.3 Part Two: GCC Specific Organizations and Plans

This section describes unique characteristics of each Geographic Combatant Command.

9.3.1 US Africa Command

Mission: United States Africa Command (USAFRICOM), in concert with other US government agencies and international partners, conducts sustained security engagement through military-to-military programs, military-sponsored activities, and other military operations as directed to promote a stable and secure African environment in support of US foreign policy. See Figure 9-1 for geographic AOR.

9.3.1.1 USAFRICOM Organizational Chart

Figure 9-2: USAFRICOM's Organizational Chart

9.3.1.2 USAFRICOM Component Forces

USAFRICOM has no assigned forces. USAFRICOM relies on the Request for Forces (RFF) process for force allocation. USAFRICOM component and Joint Forces Headquarters include:

- Combined Joint Task Force Horn of Africa (CJTF-HOA)
- US Army, Africa Command (USARAF)
- US Air Forces, Africa Command (AFAFRICA)
- US Marine Corps, Forces Africa Command (MARFORAF)
- US Naval Forces, Africa Command (NAVAF)
- US Special Operations Command, Africa (SOCAF)

9.3.1.3 USAFRICOM CONPLAN 7200-09

USAFRICOM CONPLAN 7200-09: Foreign Humanitarian Assistance Operations provides guidance for conducting FDR in the USAFRICOM AOR.

CONPLAN 7200-09 Mission Statement: When directed by the President or Secretary of Defense, the Commander of USAFRICOM conducts foreign humanitarian assistance operations in support of USAID/OFDA within the designated area. Be prepared to provide additional support as requested.

9.3.1.4 USAFRICOM Uniqueness and Typical Disasters

US Africa Command is responsible for conducting military relations with 53 African countries, including the islands of Cape Verde, Equatorial Guinea, and Sao Tome and Principe, along with the Indian Ocean islands of Comoros, Madagascar, Mauritius, and Seychelles. US Central Command maintains its traditional relationship with Egypt, though USAFRICOM coordinates with Egypt on issues relating to Africa security. The African continent has a well-established humanitarian assistance/disaster response community that can respond to earthquakes, wildfires, drought, cyclones, floods, and volcanic eruptions. The vastness of the continent, the lack of traditional infrastructure networks, the variety of languages, cultures, and delicate political systems create an environment where the US military may be asked to provide unique capabilities to assist in disaster response.

9.3.2 US Central Command

Mission: With national and international partners, US Central Command (USCENTCOM) promotes cooperation among nations, responds to crises, and deters or defeats state and non-state aggression, and supports development and, when necessary, reconstruction in order to establish the conditions for regional security, stability, and prosperity. See Figure 9-1 for geographic AOR.

9.3.2.1 USCENTCOM Organizational Chart

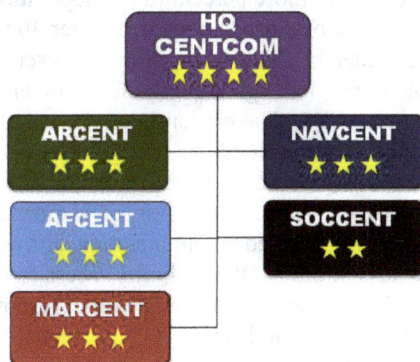

Figure 9-3: USCENTCOM's Organizational Chart

9.3.2.2 USCENTCOM Component Forces

USCENTCOM component forces include:
- US Army, Central Command (ARCENT)
- US Air Forces, Central Command (AFCENT)
- US Marine Forces, Central Command (MARCENT)
- US Naval Forces, Central Command (NAVCENT)
- US Special Operations, Central Command (SOCCENT)

9.3.2.3 USCENTCOM CONPLAN 1211-07

USCENTCOM CONPLAN 1211-07, Foreign Humanitarian Assistance/Disaster Relief, provides guidance for conducting FDR in the USCENTCOM AOR.

CONPLAN 1211-07 Mission Statement: When directed, USCENTCOM will support the lead federal agency and conduct FHA/DR operations in the AOR in order to minimize loss of human life, alleviate human suffering, and mitigate economic impact of disasters.

9.3.2.4 USCENTCOM Uniqueness and Typical Disasters

The USCENTCOM AOR is comprised of 20 countries spanning over four million square miles in three diverse sub-regions from Egypt and the Levant, to the Arabian Peninsula (including the Gulf nations), and Central and South Asia. These regions are home to a half-billion people practicing all of the world's major religions and speaking more than 18 different languages. Several countries face economic challenges and have rapidly increasing populations such as, Pakistan, Egypt, and Iran. In 12 of the 20 countries in the region, 30 or more percent of the population is between the ages of 15 and 24. In most of those countries, another 30 percent of the overall population is under 15. This youth bulge represents tomorrow's future leadership and the region's greatest challenge in terms of education, employment and expectations. The predominant natural disasters that occur in the AOR are earthquakes and flooding. The greatest challenges in responding to natural disasters in this AOR are related to its terrain. Many countries have unimproved road surfaces, limited seaports, improved runways, landing zones and limited communications infrastructure. The vastness and remoteness of the AOR also limits access to Affected States and impacts USCENTCOMs ability to provide rapid respond within the first 72-hours following a natural disaster. USG response efforts are further complicated by the fact that countries in the USCENTCOM AOR are often reluctant to request assistance from the US Government due to politics and US Foreign Affairs Policies.

9.3.3 US European Command

Mission: US European Command (USEUCOM) conducts military operations, international military engagement, interagency partnering to enhance transatlantic security, and defends United States forward. See Figure 9-1 for geographic AOR.

9.3.3.1 USEUCOM Organizational Chart

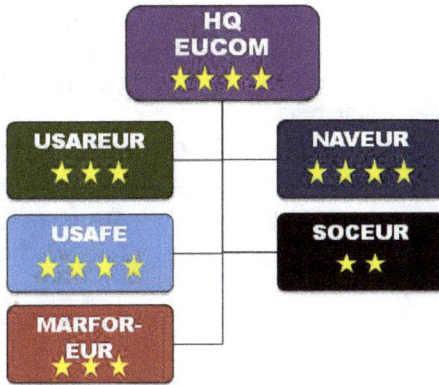

Figure 9-4: USEUCOM's Organizational Chart

9.3.3.2 USEUCOM Component Forces

USEUCOM component forces include:

- US Army, Europe (USAREUR)
- US Air Forces, Europe (USAFE)
- US Marine Forces, Europe (MARFOREUR)
- US Naval Forces, Europe (NAVEUR)
- US Special Operations Command, Europe (SOCEUR)

9.3.3.3 USEUCOM CONPLAN 4269-10

USEUCOM CONPLAN 4269-10, Foreign Humanitarian Assistance (FHA) provides guidance for conducting FDR in the USEUCOM AOR.

CONPLAN 4269-10 Mission Statement: When directed by the Secretary of Defense, the Commander USEUCOM conducts foreign humanitarian assistance operations in support of USG relief efforts in the USEUCOM AOR in order to mitigate near-term human suffering and accelerate long-term regional recovery. On order, be prepared to integrate US forces into a multinational, interagency structure for the distribution of relief aid and assistance.

9.3.3.4 USEUCOM Uniqueness and Typical Disasters

Within the USEUCOM AOR, and European Union (EU) in general, most countries have a robust and well-rehearsed capability to conduct disaster relief operations. Bi-lateral and regional agreements enhance local response efforts. As a result, DOD forces are required only for a limited number of situations. Typical disasters in the USEUCOM AOR include floods, wild-fires, and earthquakes.

9-11

9.3.4 US Northern Command

Mission: US Northern Command (USNORTHCOM) conducts homeland defense, civil support, and security cooperation to defend and secure the United States and its interests.

9.3.4.1 USNORTHCOM CONPLAN 3729

USNORTHCOM CONPLAN 3729 (DRAFT), Foreign Humanitarian Assistance and Disaster Relief (FHA/DR) in the USNORTHCOM AOR, provides guidance for the conducting FDR in the USNORTHCOM AOR.

CONPLAN 3729 Mission Statement: When directed by the President or SECDEF, US Northern Command (USNORTHCOM) conducts FHA/DR operations in support of US Government relief efforts in the USNORTHCOM AOR in order to mitigate near-term human suffering and accelerate long-term regional recovery. See Figure 9-1 for geographic AOR.

9.3.4.2 USNORTHCOM Organizational Chart

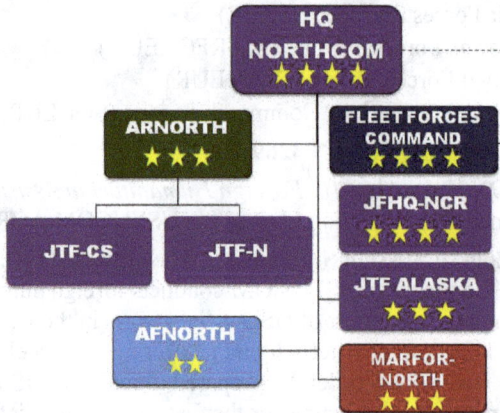

Figure 9-5: USNORTHCOM's Organizational Chart

9.3.4.3 USNORTHCOM Component Forces

USNORTHCOM component and Joint Force Headquarters include:

- US Army, North (ARNORTH)
    - Joint Task Force, Civil Support (JTF-CS)
    - Joint Task Force, North (JTF-N)
- US Air Force, North (AFNORTH/1ˢᵗ Air Force)

- US Marine Force, North (MARFORNORTH)
- Joint Force Headquarters, National Capital Region (JFHQ-NCR)
- Joint Task Force, Alaska (JTF-Alaska)

US Fleet Forces Command (USFF) is the force provider to USNORTHCOM when naval forces are required.

9.3.4.4 USNORTHCOM Uniqueness and Typical Disasters

USNORTHCOM's geographic AOR includes North America, the Gulf of Mexico, the straits of Florida; the Caribbean region inclusive of the U.S Virgin Islands, British Virgin Islands, Puerto Rico, the Bahamas and Turks and Caicos. While conducting FDR, USNORTHCOM has primarily been the supporting and not the supported Combatant Commander. USNORTHCOM and Canada Command have created a bilateral plan to conduct assistance and relief operations to one another. Mexico and the Bahamas have superb domestic response capabilities and historically, do not ask for international assistance in the wake of natural disasters.

9.3.5 US Pacific Command

Mission: US Pacific Command (USPACOM), together with other US Government agencies, protects and defends the United States, its territories, Allies, and interests; alongside allies and partners, promotes regional security and deters aggression; and, if deterrence fails, is prepared to respond to the full spectrum of military contingencies to restore Asia-Pacific stability and security.

9.3.5.1 USPACOM Organizational Chart

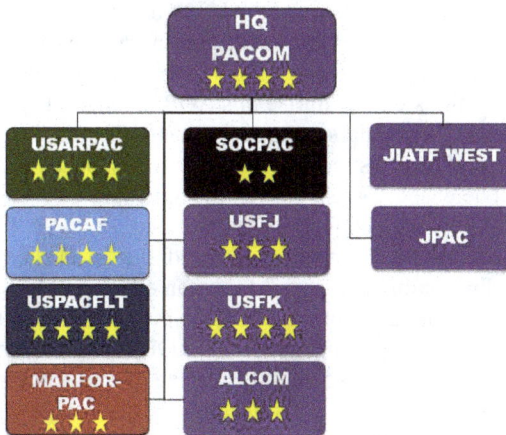

Figure 9-6: USPACOM's Organizational Chart

9-13

9.3.5.2 USPACOM Component Forces

USPACOM component and Joint Forces Headquarters include:

- US Army, Pacific (USARPAC)
- US Air Forces, Pacific (PACAF)
- US Marine Forces, Pacific (MARFORPAC)
- US Naval Forces, Pacific (COMPACFLT)
- US Special Operations Command, Pacific (SOCPAC)
- US Forces Japan (USFJ)
- US Forces Korea (USFK)
- Alaskan Command (ALCOM)
- Joint Interagency Task Force – West (JIATF West)
- Joint Prisoner of War, Missing in Action Accounting Command (JPAC)

9.3.5.3 USPACOM CONPLAN 5070-02

USPACOM CONPLAN 5070-02, Foreign Humanitarian Assistance (FDR) and Peacekeeping (PK)/Peace Enforcement (PE) Operations, provides guidance for conducting FDR in the USPACOM AOR.

CONPLAN 5070-02 Mission Statement: When directed by POTUS or SECDEF CDRUSPACOM conducts foreign humanitarian assistance/disaster relief and Peacekeeping (PK)/Peace Enforcement (PE) Operations in order to alleviate human suffering, preclude regional conflicts, and/or terminate conflicts on terms favorable to US interests.

9.3.5.4 USPACOM Uniqueness and Typical Disasters

USPACOM's AOR encompasses over half of the earth's surface, stretching from the waters off the west coast of the US to the waters west of India, and from Antarctica to the North Pole. There are few regions as culturally, socially, economically, and geo-politically diverse as the Asia-Pacific region. The 36 nations that comprise this AOR are home to more than 50 percent of the world's population, three thousand different languages, several of the world's largest militaries, and five nations allied with the US through mutual defense treaties. Two of the three largest economies are located in the AOR, along with ten of the fourteen smallest. The AOR includes the most populous nation in the world, the largest democracy, and the largest Muslim-majority nation. More than one third of Asia-Pacific nations are smaller island nations that include the smallest republic in the world and the smallest nation in Asia. This AOR is often referred to as the *Ring of Fire*, alluding to the fact that it is a zone of frequent earthquakes and volcanic eruptions that encircles the basin of the Pacific Ocean. It is shaped like a horseshoe and it is 25,000 miles long. It is associated with a nearly continuous series of oceanic trenches, island arcs, volcanic mountain ranges, and/or plate movements. 90 percent of the world's earthquakes and 81 percent of the world's largest earthquakes occur along the Ring of Fire. The Ring of Fire is a direct consequence of plate tectonics and the movement and collisions of crustal plates. Four of the ten most destructive tsunamis recorded have taken place in the PACOM AOR: two of those in the 21st Century. Cyclones and Typhoons are another typical natural disaster and nearly one-third of the world's tropical cyclones form within the western Pacific. This makes the Pacific basin the most active on Earth. Pacific typhoons have formed year round, with peak months from August to October. Along with a high storm frequency, this basin also features the most globally intense storms on record. The greatest challenge in the USPACOM AOR is the tyranny of distance. Oftentimes the vastness and remoteness of the AOR impacts USPACOM's ability to provide response efforts within the first 72-hours following a natural disaster.

9.3.6 US Southern Command

Mission: We are ready to conduct joint and combined full-spectrum military operations and support whole-of-government efforts to enhance regional security and cooperation.

9.3.6.1 USSOUTHCOM CONPLAN 6150

USSOUTHCOM CONPLAN 6150, Foreign Disaster Relief/Humanitarian Assistance, provides guidance for conducting FDR in the USSOUTHCOM AOR.

CONPLAN 6150 Mission Statement: When directed, Commander US Southern Command conducts Humanitarian Assistance/Foreign Disaster Relief operations in the USSOUTHCOM AOR in support of Affected State(s), international organizations and US lead federal agency relief efforts in order to save lives and mitigate human suffering.

9.3.6.2 USSOUTHCOM Organizational Chart

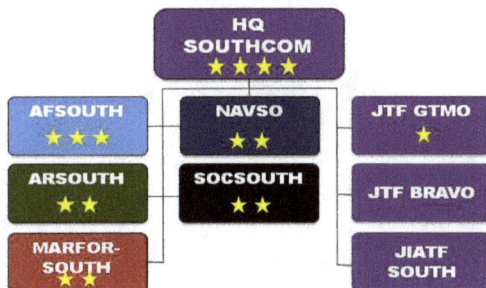

Figure 9-7: USSOUTHCOM's Organizational Chart

9.3.6.3 USSOUTHCOM Component Forces

USSOUTHCOM component forces include:
- US Army South (USARSO)
- US Air Forces Southern (AFSOUTH/12th Air Force)
- US Marine Forces, South (MARFORSOUTH)
- US Naval Forces Southern Command (NAVSO)
- US Special Operations Command South (SOCSOUTH)
- JTF Bravo (JTF-B)
- JTF Guantanamo (JTF-GTMO)
- Joint Interagency Task Force South (JIATF-SOUTH)

9.3.6.4 USSOUTHCOM Uniqueness and Typical Disasters

The countries in USSOUTHCOM's area of responsibility have deep economic ties with the US: 40 percent of all US trade stays within the Americas, ten of the seventeen US Free Trade Agreements are with countries in the Americas and 50 percent of US oil imports come from this hemisphere. In addition, two-thirds of the ships transiting the Panama Canal are going to/from US ports. While poverty is a universal problem, it is particularly problematic in the Americas. The United Nations Economic Commission for Latin America and the Caribbean report for 2008, states that extreme poverty and unequal distribution of wealth are the principle underlying causes of insecurity in the hemisphere. In this region, 33.2 percent of Latin Americans live on less than two dollars per day and 12.9 percent live in extreme poverty on less than one dollar per day. Economic conditions in the region combined with geography render some member states particularly vulnerable following a natural disaster. Typical natural disasters in USSOUTHCOM's AOR include earthquakes, floods, hurricanes, and landslides. In spite of the region's economic standing, this region is often reluctant to seek assistance from the USG in response to a disaster.

CHAPTER 10 - DEPARTMENT OF DEFENSE CROSS-CUTTING ORGANIZATIONS

10.1 Overview

This chapter describes Department of Defense (DOD) cross-cutting organizations that have significant roles in FDR operations. It includes the roles and missions of the Specified and Functional commands and DOD agencies that provide support to the Geographic Combatant Commands.

10.2 Unified Combatant Commands (Functional)

In the Unified Combatant Command (UCC) architecture, Functional Combatant Commands (FCC) serve as force and resource providers to Geographic Combatant Commanders (GCC). FCC have global response authority and often serve in a Supporting Command capacity to the GCC.

The FCC are:
- US Special Operations Command (USSOCOM)
- US Strategic Command (USSTRATCOM)
- US Transportation Command (USTRANSCOM)

NOTE: *US Joint Forces Command (USJFCOM) is scheduled for disestablishment effective 31 August 2011. Many of its functions will revert to the Services or the Joint Staff; however, as of the publication date, final disposition of all functional elements has not been determined. One JFCOM agency pending disposition is the Joint Personnel Recovery Center described in Section 9.2.8.*

10.3 US Special Operations Command
10.3.1 Mission

Special Operations Forces (SOF) are characterized by small units of specially trained and selected personnel that are capable of conducting high-risk missions in politically sensitive environments. SOF have the ability to deploy rapidly and operate effectively in austere environments with little or no infrastructure. Those characteristics make them ideally suited to respond to a disaster in an area of limited accessibility, infrastructure, or some other variable that hinders the ability of NGOs and other response entities to access and support an affected population. SOF assets may operate under USSOCOM or a GCC in the form of a Theater Special Operations Command (TSOC).

In addition to these baseline capabilities associated with SOF, the following capabilities support humanitarian assistance and disaster response missions across all GCCs:

- 95th Civil Affairs Brigade
- Combat Controller Teams
- Military Information Support Operations

10.3.2 95th Civil Affairs Brigade

Assigned to USSOCOM, the 95th Civil Affairs (CA) Brigade, stationed at Fort Bragg, NC is currently the Army's only Active duty CA Brigade (for more information on conventional forces CA missions and assets, see chapter 11). The 95th CA Brigade consists of 1200 CA generalists, and provides global, rapid deployment capabilities in support of FDR operations typically for a maximum duration of six months or until relieved by Reserve Component CA units.

10.3.3 Combat Controller Teams

Combat Controller Teams (CCT) are organized, trained, and equipped to rapidly establish and control the air-ground interface. Functions include landing zone assessment and establishment; air traffic control; command and control communications; and removal of obstacles with demolitions. CCTs provide a unique capability and deploy with joint air and ground forces for humanitarian assistance and austere airfield operations.

10.3.4 Military Information Support Operations

Military Information Support Operations (MISO) personnel provide the commander with the ability to communicate information to large audiences via radio, television, leaflets, and loudspeakers. MISO personnel language skills, regional orientation, and knowledge of communications media provide a means of delivering critical information to disaster victims. In addition to supporting military units, MISO personnel provide interagency support and can be a useful asset in strategic communications during FDR operations.

10.4 US Strategic Command

US Strategic Command's (USSTRATCOM) mission is strategic defense of the United States. USSTRATCOM may be called upon to provide geospatial-tracking assets in support of FDR operations.

10.5 US Transportation Command

The mission of the US Transportation Command (USTRANSCOM) is to develop and direct the Joint Deployment and Distribution Enterprise to globally project strategic national security capabilities; accurately sense the operating environment; provide end-to-end distribution process visibility;

10-2

and responsive support of Joint, USG and Secretary of Defense-approved multinational and non-governmental logistical requirements. In fulfilling this mission, USTRANSCOM provides critical core competencies in the USG's ability to conduct humanitarian assistance and disaster response operations.

USTRANSCOM integrates transportation and distribution across the range of FDR requirements through its three service component commands:

- Military Surface Deployment and Distribution Command (SDDC) is the Army component
- Military Sealift Command (MSC) is the Navy component
- Air Mobility Command (AMC) is the Air Force component

Together, those organizations make available transportation and logistics capabilities to the supported GCC, the JTF Commander, and USG agencies that assist in relieving the burden on the Affected States during FDR operations.

10.5.1 Joint Deployment and Distribution Operations Center

Each GCC operates a Joint Deployment and Distribution Operations Center (JDDOC), which directs, coordinates and synchronizes assigned and attached forces' deployment and redeployment execution, and distribution operations to enhance the GCC's ability to execute logistics plans with support of national partners. The JDDOC works closely with USTRANSCOM, theater Service components, and other national partners to influence the distribution process from the theater's strategic port of entry to the point of need.

10.5.2 JTF Port Opening

A key early step in FDR operations is to assess, and when necessary, open aerial and sea ports. The FDR response may involve unique requirements for getting supplies and personnel in and out of the Affected State. USTRANSCOM has developed a joint enabling capability called Joint Task Force - Port Opening (JTF-PO) to open Aerial Ports of Debarkation (APOD) and Seaports of Debarkation (SPOD). Staffing for those organizations is provided by AMC, MSC, and SDDC. JTF-POs joint expeditionary capabilities can rapidly establish, initially operate, and clear a port of debarkation and conduct cargo handling and movement operations to a forward node, facilitating port throughput in support of GCC-executed contingencies.

The primary JTF-PO missions are early port opening and clearing of cargo out to 10km to prevent a backlog at a vital reception point. Based on the situation, jointly trained JTF-PO forces can be deployed quickly and are designed to remain in theater for 45-60 days.

The GCC may have other port opening capabilities resident within theater Service components. The Army and Navy have expertise and equipment for seaport opening operations, including cargo handling, engineering, beachmasters, salvage and dive units to provide a wide range of supply and transportation support critical for FDR missions.

For aerial port opening, theater Air Force components may also have similar JTF-PO capabilities. If a JTF-PO is established to evaluate and reopen key transportation nodes, the organizations listed below can be combined in several iterations to provide the JTF Commander a tailored solution to meet the requirements of an Affected State.

10.5.2.1 Contingency Response Groups

The US Air Force's Contingency Response Groups (CRG) are responsible for assessing and determining the aerial ports infrastructure capabilities within an Affected State. Once an airport has been cleared for operations, the CRG will remain in place to assist the Affected State.

Contingency Response Element (CRE) is an organization that provides global air mobility support where previously it was insufficient or non-existent. A CRE provides continuous on-site management of Air Mobility Command (AMC) airfield operations including command and control, aerial port services, maintenance, security, weather, and intelligence. Those critical elements needed to ensure safe, efficient airfield operations for all operations. A CRE is composed of mission support elements from various unit type codes and AMC Quick Reaction Force for mobility operations on both planned and short notice missions.

As the air component of the USTRANSCOM, AMC enables forces to rapidly mobilize and deploy to support the Combatant Commanders course of action to meet national objectives. To manage, coordinate, and control that air mobility mission, AMC established a global air mobility support structure of fixed and deployable elements. One of the key deployable elements of this support structure is the CRE.

The CRE is a short-term, composite, deployed organization consisting of command and control and essential mission support elements. The organizational structure of a CRE is the skeleton of a typical mobility wing, with the CRE providing command and operations functions, deployed maintenance, and aerial port functions. This compact force deploys to locations where a fixed AMC C2 and support structure is limited or nonexistent. The CRE is yet another mobile arm of AMC's Global Air Mobility Support System. The CRE provides the command and control functions required to support AMC's worldwide mobility operations. CREs normally conduct autonomous operations from austere locations but can also augment the infrastructure at established civilian or military airfields. They provide minimum cargo loading and a quick turn en route AMC mission support during mobility operations. The following additional organizations may deploy to support the AMC global reach mission:

Contingency Response Team (CRT) performs the same functions as a CRE, but for a more limited flow of aircraft and normally a shorter duration.

Communication Support Team (CST) is part of the communication segment of the CRE, but are capable of employing a communications suite independent of the CRE.

Airfield Survey Team (AST) is responsible for conducting airfield surveys worldwide. The AST Team Leader is certified in conducting these specialized surveys. The composition of a team is determined by the team chief tasked to complete the survey. A typical survey team includes: civil engineering; security; airfield operations; petroleum, oil, and lubricants (POL) representative; and aerial port personnel.

Tactical Operations Center (TOC) is the nerve center at each CRE and CRT location. The TOC supports mission planning and scheduling, and gives the CRE and CRT the ability to track the status of resources and manage passenger and cargo operations. The TOC offers a variety of communications and information processing systems to link the TACC and a CRE/CRT. Similar to a base operations center, the TOC can relay mission essential information from anywhere in the world.

10.5.2.2 Rapid Port Opening Elements

The US Army's Rapid Port Opening Elements (RPOE) provide surface transportation expertise for JTF-PO. Their capabilities include receiving and transloading cargo as an initial port opening element.

The addition of three RPOE to the Army's Military Surface Deployment and Distribution Command brings an expeditionary answer to the challenge of logistics support in contingency response operations for the DOD. As the surface piece of USTRANSCOM's JTF-PO, RPOE deploy as part of a joint expeditionary logistics force to establish a port of debarkation and forward distribution node. The RPOE provides in-transit visibility and conduct clearance and distribution operations. They also receive and transload cargo as an initial-entry port opening force until relieved by - or are integrated into - follow-on sustainment forces. Initially conceived to support aerial port of debarkation operations, the capabilities of the RPOE is expanding to include seaport of debarkation capability.

UNCLASSIFIED

10.5.2.3 Expeditionary Port Units

Expeditionary Port Units (EPU) are the Navy component of the JTF-PO. They are primarily concerned with ensuring the stability and safety of seaport operations. Their expertise allows the JTF Commander to determine the location where waterborne supplies and personnel arrive.

10.5.3 Global Standing Joint Force Headquarters

Effective 30 June 2011, the Joint Enabling Capabilities Command (JECC) was reassigned from US Joint Forces Command (USJFCOM) to USTRANSCOM as the Global Standing Joint Force Headquarters (GSJFHQ).

> **NOTE:** *In accordance with Resource Management Decision 700A3 the Secretary of Defense directed the disestablishment of the six GCC Standing Joint Force Headquarters (SJFHQ). The effective date for disestablishment of the SJFHQ is 30 September 2011. It also directs the Chairman of the Joint Chiefs of Staff to establish a GSJFHQ by transitioning the JECC and using the resources from disestablishment of USJFCOM.*

Currently, the GSJFHQ is comprised of one *Ready-Package* capable of providing immediate, short-duration support to establish, organize and operate joint force headquarters across the globe and combines capabilities from the six unique functional areas of plans, operations, logistics, information superiority/knowledge management, communications and public affairs. A second *Ready-Package* will be established effective 1 October 2012 (see figure 10-1).

USTRANSCOM's GSJFHQ (JECC) provides joint force commanders with distinct capabilities that offer short duration support to establish, organize and operate a joint force headquarters. The GSJFHQ is responsible for providing trained, responsive, and cohesive capabilities to assist in rapid establishment of a joint task force headquarters (JTF HQ).

The GSJFHQ's mission is to employ, manage, and develop existing Joint Enabling Capabilities (JEC). This includes the following deployable support elements: the Joint Deployable Team (JDT), the Joint Communications Support Element (JCSE), and the Joint Public Affairs Support Element (JPASE).

JECC Transition from JFCOM to TRANSCOM
Global Standing Joint Force Headquarters

Figure 10-1: JECC Transition to GSJFHQ

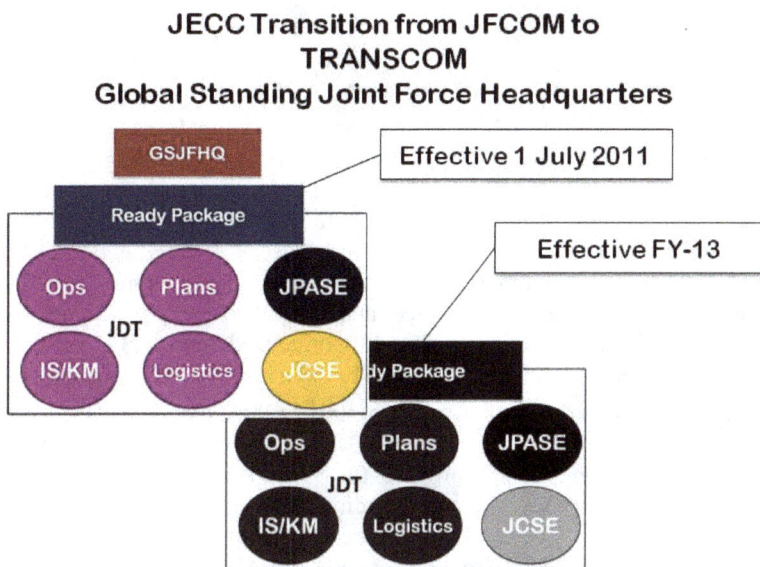

10.5.3.1 Joint Deployable Team

The JDT is a flexible employment package composed of personnel from the four JECs of operations, plans, knowledge/information management, and logistics. Teams enable a JTF HQ to rapidly form, plan, operate, and integrate with interagency and non-military elements.

10.5.3.2 Joint Communications Support Element

The JCSE can rapidly deliver secure, reliable, and scalable command, control, communications, and computer capabilities ranging from small mobile teams to full-sized JTF HQ deployments.

10.5.3.3 Joint Public Affairs Support Element

The JPASE provides the joint force commander with a trained, equipped, scalable, and expeditionary joint public affairs capability supporting worldwide operational requirements.

10.5.4 Joint Logistics Over-The-Shore

Joint Logistics Over-The-Shore (JLOTS) provides a ship discharge capability without the benefit of deep draft-capable port facilities, such as in austere locations, to augment existing port facilities and damaged ports. JLOTS occurs when Navy and Army Logistics Over-The-Shore (LOTS) forces conduct operations together. The scope of JLOTS operations extends

10-7

from acceptance of ships for off-load, through arrival of equipment and cargo at inland staging and marshalling areas. GCCs have overall responsibility for JLOTS operations in their AORs.

During the first days of FDR operations, a JTF-PO Joint Assessment Team (JAT) may determine that JLOTS is required. USTRANSCOM provides subject matter experts (SME) and JLOTS planning support to the supported GCC staff for concept development. Specific JLOTS operations will be identified by the JTF Commander including tentative JLOTS sites and force requirements. The supported GCC can then request JLOTS forces to begin expeditious movement of ships, forces and equipment. Forces assigned to conduct the JLOTS operation are normally organized under a JLOTS Commander operating in support of a JTF.

USTRANSCOM is responsible for orchestrating the movement of anticipated sealift assets. The USTRANSCOM Commander, through his component command, SDDC, provides the Single Port Manager (SPM) for worldwide common-user seaports, including those discharge sites requiring JLOTS capabilities. The strategic sealift employed in support of JLOTS operations includes MSC common-user ships, US Maritime Administration (MARAD) maintained vessels (referred to as the Ready Reserve Force [RRF]), and pre-positioning ships. RRF ships are maintained in a reduced operating status and require the USTRANSCOM Commander to authorize their activation. For more detail see JP 4-01.6, *Joint Logistics Over-the-Shore [JLOTS]*.

10.5.5 Global Patient Movement Requirements Center

USTRANSCOM's Global Patient Movement Requirements Center (GPMRC) oversees and tracks patient movement through the USTRANSCOM Regulating and Command and Control Evacuation System (TRAC2ES) in coordination with GCCs.

10.5.6 USTRANSCOM Fusion Center

USTRANSCOM operates a fusion center that merges the headquarters, as well as that of each Transportation Component Command (TCC), acting as one team to optimize, synchronize, and manage the movement of global requirements. The Center is augmented, as required, to support increased operations tempo and to ensure the customer is supported in the most expeditious manner possible.

10.5.7 Subject Matter Experts

USTRANSCOM provides distribution subject matter experts (SME) to augment GCC JDDOC. Those SME also function as liaisons back to the USTRANSCOM Fusion Center facilitating the movement of personnel and materiel into and out of the Area of Operations (AO). USTRANSCOM

10-8

shall coordinate with the Lead Federal Agency (LFA) and keep them apprised of all actions in the AO.

10.5.8 Contracting

Operational Contract Support provides access to commercial assets and services and uses local providers to reduce reliance on extended supply chains. Using contracts to procure Affected State capabilities, or those from commercial providers that are familiar with the AO, helps to stabilize the operation by presenting a familiar face to the populace and reducing what some see as a threatening military presence. Contracting allows other USTRANSCOM personnel to focus their efforts on missions that do not readily lend themselves to contract support. It also reduces the number and types of operations that need to be transitioned to other agencies once relief efforts are concluded.

USTRANSCOM contracting SMEs are an integral part of the initial support personnel. From the outset of the response, they coordinate their activities closely with the GCC-designated lead service for contracting (or comparable organization) to ensure a well-synchronized contracting effort within the operational area.

10.6 DOD Agencies

The following DOD Agencies typically support FDR operations:

10.6.1 Defense Logistics Agency

Defense Logistics Agency (DLA) provides the Services, other federal agencies, and joint and allied forces with a variety of logistics, acquisition, and technical services. DLA provides nearly all consumable items America's military forces need to operate, including food, fuel and energy, uniforms, medical supplies, and construction and barrier equipment. In addition, DLA manages the reutilization of military equipment, provides catalogs and other logistics information products, and offers document automation and production services. In fulfilling these missions, DLA plays a role in the provision of humanitarian aid in times of crisis. See link for DLA customer handbook/publications:

https://headquarters.dla.mil/DLA_Customer/Operations/Publications.aspx

10.6.1.1 DLA Contingency Support Team

The DLA Contingency Support Team (DCST) provides integrated supply management support for all DLA-managed materiel required in-theater. DCST determines the availability and source of supplies, provides guidance on all DOD and DLA regulations, and procedures regarding DLA commodities. Additionally, DCST can authorize deviations from regulations as appropriate for in-theater operations, provide in-transit visibility of DLA items, and coordinate packaging, shipping, movement, and priority of DLA items. A useful website used by DCSTs is:www.logtool.net

10.6.1.2 Expeditionary Disposal Remediation Team

The purpose of the Expeditionary Disposal Remediation Team (EDRT) is to manage and operate the disposal function, to include hazardous waste disposal required in-theater. EDRT can negotiate disposal and resale with Affected States. Additionally, EDRT provides storage and care of theater, DLA Disposition Services field office (formerly called DRMO) assets. Finally, EDRT provides demilitarization, downgrading of material to scrap, cannibalization, and sale or donation of surplus property.

10.6.1.3 Fuel Support Team

The Fuel Support Team (FST) serves as the liaison between DLA (Defense Energy Supply Center), the Defense Fuel Regions, and theater Joint Petroleum Offices. FST also provides quality assurance for petroleum matters, and conducts quality surveillance of US bulk fuel.

10.6.2 Defense Security Cooperation Agency

Defense Security Cooperation Agency (DSCA) is part of OSD Policy (OSD/P) and, though not responsible for creating policy, DSCA plays a significant role in implementing policy and ensuring its adherence.

10.6.2.1 Humanitarian Assistance, Disaster Relief, and Mine Action

Within DSCA, the Programs Office of Humanitarian Assistance, Disaster Relief, and Mine Action (DSCA/PGM/HDM) provides program management and execution oversight of the DOD humanitarian and mine action assistance activities conducted by the GCC. DSCA/PGM/HDM also manages the Overseas Humanitarian Disaster and Civic Aid (OHDACA) appropriation and its funded activities, to include, the Humanitarian Assistance (HA), Excess Property (EP), Humanitarian Mine Action (HMA), Denton (Space Available) and Funded Transportation programs, and Foreign Disaster Relief/Emergency Response (FDR/ER) activities as directed by SECDEF.

Functions accomplished by HDM include:

- Providing direction and control of the activities of the Humanitarian Demining Training Center (HDTC)
- Overseeing GCC mine action activities
- Overseeing Foreign Disaster Relief and Emergency Response activities
- Overseeing the DOD Humanitarian Assistance Program (HAP)
- Managing the OHDACA appropriation

- Managing and developing the Overseas Humanitarian Assistance Shared Information System

DSCA manages and coordinates the funding of FDR activities with the interagency and DOD components. DSCA works with other OSD policy offices, Department of State, USAID/OFDA, and DOD components to facilitate and manage FDR activities, and is the lead for coordinating with the GCC and USAID/OFDA on any excess property or humanitarian daily ration requirements for the affected area.

DSCA's role in FDR includes assisting in capturing cost, reimbursement, and resourcing GCCs for the mission.

The following humanitarian and mine action programs established under the legal authorities cited govern the use of OHDACA appropriated funds:

- 10 United States Code (USC) § 404 *Foreign Disaster Assistance,* resources DOD's participation in foreign disaster relief missions; procures, manages, and arranges for delivery of humanitarian daily rations where required.
- 10 USC § 2557 *Excess nonlethal supplies: availability for homeless veteran initiatives and humanitarian relief,* provides excess, non-lethal, excess property to authorized recipients, and arranges for its transportation.
- 10 USC § 2561 *Humanitarian Assistance Act,* establishes The Funded Transportation Program, permits transportation of cargo and DOD non-lethal excess property worldwide for non-governmental/international organizations (NGO/IO). This authority provides for the actual cost of transportation and the payment of any associated administrative costs incurred.
- 10 USC § 402 *The Denton Transportation Program* arranges for no-cost, space-available transportation of NGO donors for delivery of humanitarian goods to countries in need.
- 10 USC § 407 *The Humanitarian Mine Action (HMA) Program,* assists countries that are experiencing the adverse affects of uncleared landmines and other explosive remnants of war. The program is managed by the combatant commanders and contributes to unit and individual readiness by providing unique in-country training opportunities that cannot be duplicated in the United States. The Humanitarian Demining Training center (HDTC), located at Fort Leonard Wood, Missouri, provides training to deploying forces and others on mine action topics and provides technical assistance to the GCCs.

10.6.3 National Geospatial-Intelligence Agency

Geographic orientation of a disaster stricken state is key for the JTF Commander to understand early and routinely throughout FDR operations. National Geospatial Intelligence Agency (NGA) provides the JTF Commander with geographic information regarding Affected States following disasters. Many organizations can provide geospatial imagery including open source material via websites such as ReliefWeb. The JTF Commander typically receives NGA information from the GCC's Geospatial Planning Cell (see Section 9.2.10), which is capable of providing diverse and detailed imagery of Affected States, including pre-disaster views, affected areas, and key terrain features. The release of NGA materials should be coordinated with NGA via the JTF Foreign Disclosure Officer.

10.6.4 Defense Information Systems Agency

Defense Information Systems Agency (DISA) is a combat support agency that engineers and provides command and control capabilities and enterprise infrastructure to continuously operate the global information grid (GIG) and assure a global, net-centric enterprise in direct support to joint warfighters, national-level leaders, and other mission and coalition partners across the full spectrum of global operations.

DISA maintains valuable resources in the form of Combatant Command Field Offices, which are equipped to understanding the nuances of each region of the globe as well as the GCC mission and other regional dynamics. The DISA Field Offices are:

- DISA AFRICOM - Contact info: Mohringen, Germany; DSN (314) 421-2979; Commercial 49(0) 711-729-2979

- DISA CENT - Contact info: MacDill AFB, FL; DSN 651-6403 Commercial (813) 827-6403

- DISA CONUS - Contact info: Scott AFB, IL; DSN 770-8840 / 8801; Commercial (618) 220-8840 / 8801

- DISA EUROPE - Contact info: Enterprise Services Manager; DSN: 314-434-5646; Commercial 49 (0)711-68639 5646

- DISA NORTHCOM - Contact info: Peterson AFB, CO; DSN 692-3800; Commercial (719) 554-3800 / 5962

- DISA PACIFIC - Contact info: Ford Island, HI; DSN (315) 472-0051; Commercial(808) 472-0051; or Wheeler Army Airfield, HI; DSN (315) 456-1665/1647; Commercial (808) 656-1665/1647

- DISA SOUTHCOM - Contact info: Miami, FL; DSN 567-1671; Commercial (305) 437-1671

DISA has a unique role in FDR operations in providing the JTF with information on GCC-unique systems. As DOD's satellite communications leader, DISA uses the Defense Satellite Communications System to provide frequency and bandwidth support to all organizations in FDR operations. That support includes super high frequency missions that provide bandwidth for Navy ships and Marine Expeditionary Units participating in FDR operations. This also has included handling satellite communications for the US Air Force. In the event that an Affected State requires an air traffic control capability, DISA may provide satellite communications for the US Air Force. DISA is also capable of providing military, ultra-high frequency channels and contracting for additional commercial satellite communications that may increase the capabilities of FDR operations.

DISA owns the Transnational Information Sharing Cooperation (TISC) tool, which allows NGO, other nations, and US forces to track, coordinate, and better organize FDR efforts. The All Partners Access Network (APAN) portal allows any FDR organization to sign on and collaborate, using APAN's Web-based social networking services. Through TISC, users can coordinate and efficiently direct FDR supplies arriving by air and ship, as well as rescue and medical experts arriving to assist the Affected State.

This page intentionally left blank

CHAPTER 11 - DEPARTMENT OF DEFENSE TACTICAL RESPONSE ORGANIZATIONS IN FDR OPERATIONS

11.1 Overview

The primary mission of the Department of Defense (DOD) is national defense. Department of Defense Directive (DODD) 5100.01, *Functions of the Department of Defense and Its Major Components*, and establishes the broad functions of the Department and its major Components. The core mission areas of the Armed Forces include military operations and activities required to achieve the security objectives addressed in the National Security Strategy, National Defense Strategy and the National Military strategy. As prescribed by higher authority, the Department of Defense shall maintain and use armed forces to:

- Support and defend the Constitution of the United States against all enemies, foreign and domestic.
- Ensure, by timely and effective military action, the security of the United States, its possessions, and areas vital to its interest.
- Uphold and advance the national policies and interests of the United States.

Disaster relief, under the broader mission of Stability Operations is a directed mission for DOD.

> *A core responsibility of the US Government (USG) is to protect the American people –in the words of the framers of our Constitution, to 'provide for the common defense'."*
> *National Defense Strategy*
> *June 2008*
>
> *"We must be prepared to support and facilitate the response of the United States Agency for International Development and other USG agencies to humanitarian crises."*
> *National Military Strategy*
> *February 2011*

At the direction of the Secretary of Defense and with the concurrence of the Secretary of State, DOD organizations participate in FDR operations. The majority of requests involve tactical level capabilities unique to DOD. Foreign Disaster Relief (FDR) is inherently a logistics based operation.

As such, request for DOD FDR support typically include intra-theater airlift, supply chain management, logistics distribution, medical logistics, aerial and ground transportation, engineering, communications, and civil affairs capabilities and assets. On a few occasions, DOD has been asked to temporarily augment the Affected State's existing Health Affairs and Security/Force Protection capabilities. This chapter provides an overview of those unique DOD tactical capabilities.

> **NOTE:** *Tactical organizations and units are often named in relation to their wartime functions, as such, regardless of their potential utility in FDR operations, unit nomenclatures have not been altered in this chapter to remove references to combat functions.*

11.2 Logistics

DOD's ability to rapidly distribute supplies, house personnel, and transport significant quantities of supplies, personnel and equipment into isolated areas with limited infrastructure can be a critical enabler to the success of disaster relief operations. DOD's unique tactical logistics capabilities include distribution of supplies and equipment (such as food, water, fuel, construction materials, and other basic life sustainment items); engineering (including infrastructure assessments, inspection of facilities utilities repairs, rapid repair of roads and bridges and base camp support); aerial, ground and waterborne transportation; and, material handling equipment (to off-load supplies and equipment). These logistics assets are capable of filling critical gaps in the humanitarian response in support of the Affected State and the international humanitarian community.

> **NOTE:** *During FDR operations, USAID/OFDA prefers to operate under a pull system (where mission priorities are based upon commodities requested) versus a push system (where items are shipped, based on availability, without regard for specific need or request). Failure to use a pull system can delay the delivery and distribution of critical relief commodities.*

11.3 Distribution

The distribution of humanitarian aid and disaster relief supplies must be coordinated in conjunction with the Affected State's overarching response and recovery plan. The United States Transportation Command

11-2

(USTRANSCOM) Joint Deployment Distribution Operations Center (JDDOC) and theater JDDOC work together to develop the overarching distribution plan from the strategic to tactical level. Furthermore, DOD coordination with US Agency for International Development/Office for US Foreign Disaster Assistance (USAID/OFDA), and the World Food Program, the lead for the United Nations Logistics Cluster, will reduce congestion and facilitate the efficient flow of supplies, equipment, and personnel to the Affected State.

> **NOTE:** *A distribution point established at the seaport or aerial port of debarkation is essential to reducing congestion and facilitating the efficient flow of supplies and personnel to the Affected State. Initially, the JTF Theater distribution point should be coordinated by Joint Task Force- Port Opening (JTF-PO) personnel (if present), or the JDDOC and in collaboration with USAID/OFDA, or the logistics cluster and the Affected State's lead agency.*

11.4 Ground Transportation

The US Army and Marine Corps have the preponderance of rapidly deployable ground transportation assets, although Navy assets may also be available. Those ground transportation units include medium and heavy lift transport trucks; material handling equipment, such as forklifts and rough terrain cargo handling systems; and, experienced logisticians to facilitate rapid loading, off-loading, transfer, and distribution of relief supplies and equipment.

11.5 Aerial Transportation

The Joint Task Force (JTF), Joint Forces Air Component Commander (JFACC), will perform management of DOD aerial transport or flight operations. JFACC's role is described in Section 9.2.8. DOD aviation assets typically allocate the preponderance of available aerial transportation support to disaster relief operations, the remainder of its available support will be in direct support of deployed DOD forces. Aviation units supporting FDR will routinely provide Aviation Liaison Officers (ALO) to coordinate DOD flight operations with USAID/OFDA/DART, the Affected State, or, in some cases, with the UN Logistics Cluster.

11.5.1 Aviation Request and Assignment Process

Requests for DOD aviation support should be submitted through the USAID/OFDA or the Disaster Assistance Response Team (DART) using the Mission Tasking Matrix (MiTaM) process. (See Appendix C for MiTaM format). When an organization requests support they must identify the specific parameters required (e.g., cargo or personnel to be transported, delivery timeline, originating location, and destination), USAID/OFDA then validates the mission taskings using the MiTaM validation process. The supporting agency is responsible for resourcing the requirement and providing the appropriate air asset to accomplish the mission.

11.5.2 DOD Aviation Support

DOD FDR aviation mission support may include:
- Intra-theater airlift of cargo, palletized loads and personnel
- Aeromedical Evacuation (AE)
- Aerial logistics distribution
- Medical evacuation (MEDEVAC)
- Aerial assessments/photography
- Personnel transport
- Population evacuation
- Firefighting (bucket operations)
- Airfield opening
- Airspace management
- Command and Control
- Search and Rescue
- Personnel Recovery

11.5.3 US Air Force Aviation Capabilities

11.5.3.1 Intra-theater Airlift

If authorized by the Secretary of Defense, the US Air Force (USAF), through USTRANSCOM and Air Mobility Command, is capable of providing intra-theater, fixed-wing aircraft support. The USAF primarily supports FDR using large cargo aircraft, including the C-130, C-17, and C-5 aircraft. While individual aircraft capabilities may vary all are capable of transporting oversized and palletized relief items often too large for commercial aircraft. The use of C-130, C-17 and C-5 aircraft requires runways greater than 3,500 feet in length.

11.5.3.2 Aeromedical Evacuation

Aeromedical Evacuation (AE) units evacuate patients under the care of crewmembers and Critical Care Air Transport Teams (CCATT), using

11-4

fixed-wing aircraft. AE units transport sick, seriously ill, or injured military personnel and their dependents stationed outside the continental United States (OCONUS). Under certain conditions, and if approved by the Secretary of Defense AE units can also transport seriously ill or injured civilians.

The C-130 is the predominant AE platform, but it may be supplemented by the C-17; capacities for each platform are:

- C-130 - 74 litters or 92 ambulatory patients and attendants
- C-17 - 60 litters and 54 ambulatory patients and attendants

11.5.3.3 Unmanned Aerial Systems

USAF Unmanned Aerial Systems (UAS) such as Global Hawk can operate far above normal commercial traffic, while providing situational assessment to ground commanders. Intermediate systems, such as the Predator, have supported recent disaster operations. These platforms can provide aerial imagery and full-motion video capability and dramatically increase real-time situational awareness for the (JTF) Commander.

Figure 11-1: MQ-1 Unmanned Aerial Vehicle

11.5.4 US Army Aviation Capabilities

The Army is capable of providing a mix of fixed and rotary-wing capabilities to the JTF Commander and dependent upon the nature of the disaster and request, an Aviation Task Force would typically deploy. The Army's fixed wing fleet consists of C-12 and UC-35 aircraft. These aircraft are capable of transporting limited quantities of personnel, and cargo. These aircraft would typically be used in support of DOD, GCC, and JTF leadership transport during FDR operations.

Helicopters are critical assets during FDR operations. When roads, bridges, and railroads may have been damaged by disaster, helicopters may be the only method to deliver relief supplies and transport relief workers to the operational area. Depending on the scale of the disaster and Affected State

11-5

organic capabilities, an Aviation Task Force would normally deploy in support of FDR operations.

11.5.4.1 Army Combat Aviation Brigade

An Army Aviation Task Force typically includes a Combat Aviation Brigade (CAB) headquarters for command and control of assigned aircraft. A CAB is capable of tailoring task forces to mission requirements. A CAB has six organic battalions, consisting of utility, cargo, and observation aircraft, including UH-60 medium lift; CH-47 heavy lift; OH-58 series observation; and Unmanned Aerial Systems (UAS) aircraft. CABs can deploy as a brigade, a large composite Aviation Task Force or small tailored force package. CABs are capable of conducting the following missions:

- Air movement of equipment and supplies, including joint logistics-over-the-shore and aerial transport; logistics resupply (internal and external loads); and personnel transport
- UH-60 medical evacuation (MEDEVAC) equipped aircraft and staffed for patient transport (with en route care)
- CH-47 and UH-60 are capable of performing casualty evacuation (patient transport without en route care)
- CH-47 aircraft are also capable of high-altitude operations and oversized heavy and special equipment movement
- Insertion and extraction operations
- Aerial assessments of key infrastructure; aerial photography, of roads, bridges, airports, helipads, landing strips and seaports with all aircraft as well as full motion video using OH-58 helicopters and Army UAS
- Aviation logistics and maintenance support
- Forward Arming and Refueling Point (FARP) operations; providing fuel where and when needed
- Airspace Command and Control Cells maintain a real-time single, integrated picture of airspace
- Air Traffic Service/Airfield Operations (deconflicting, synchronizing, and integrating all airspace requirements, including UAS)

C-12	**UC-35**

HH-60 MEDEVAC	**UH-60 BLACK HAWK**
CH-47 CHINOOK	**OH-58D KIOWA WARRIOR**
SHADOW UAS	**HUNTER UAS**

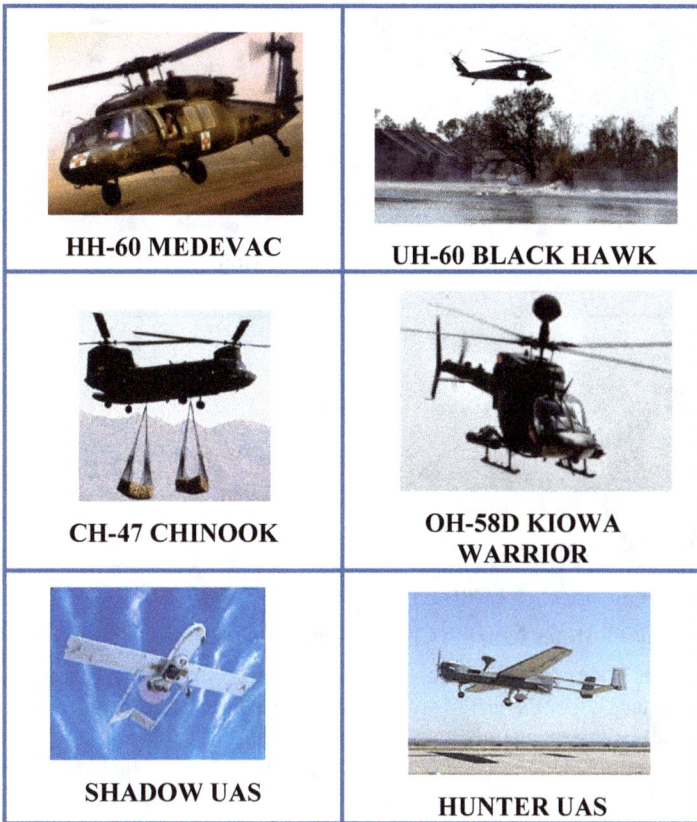

Figure 11-2: US Army Aircraft

11.5.5 United States Navy Aviation Capabilities

United States Navy (USN) support capabilities include helicopter and maritime patrol aircraft operations. P-3C, Orion, aircraft are capable of providing aerial video and photography that can be transmitted, to the JTF. USN rotary-wing assets include the H-53 heavy lift and H-60 medium lift fleets. All helicopters are capable of aerial search, internal and external cargo movement, personnel transfer, and personnel recovery by hoist from land and water areas that do not permit landing. Heavy lift helicopters, specifically MH-53E, also tow sonar equipment can be used to determine if waterways are clear prior to reopening to vessel traffic.

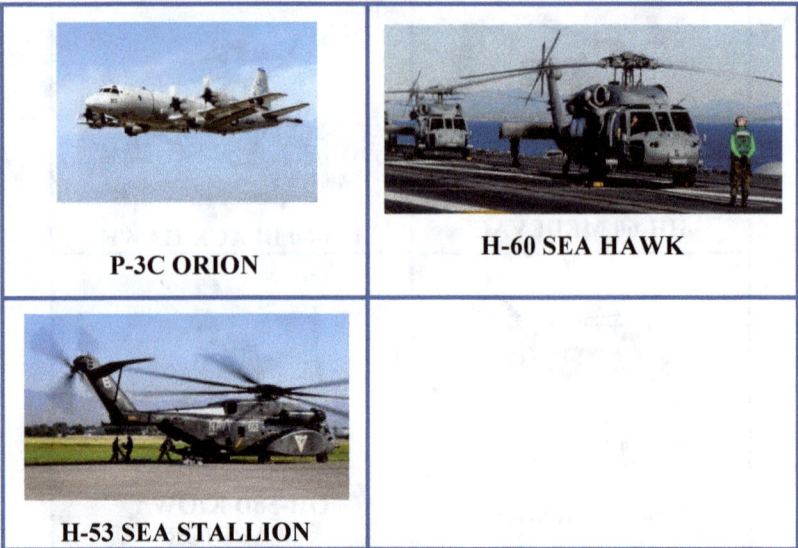

P-3C ORION

H-60 SEA HAWK

H-53 SEA STALLION

Figure 11-3: US Navy Aircraft

11.5.6 United States Marine Corps Aviation Capabilities

The United States Marine Corps (USMC) provides support to FDR by using expeditionary capabilities to respond. Marine Forces plan for the use of forces; advise on the proper employment of USMC forces; and coordinates with and support USMC forces within the AOR. The Marine Corps UH-1 provides a command and control platform, while the CH-46 and CH-53 are capable of conducting medium to heavy lift transport. The MV-22 tilt-rotor aircraft offers a unique combination of speed and landing versatility.

Typically, in a FDR environment, if tasked, the USMC will provide a Marine Air Ground Task Force (MAGTF). A MAGTF is comprised of four core elements: a Command Element (CE), a Ground Combat Element (GCE), an Aviation Combat Element (ACE), and a Logistics Combat Element (LCE). As a modular organization, the MAGTF is tailorable to each mission through task organization. If tasked, the USMC is capable of supporting FDR operations using fixed-wing (C-130), tilt-rotor aircraft (MV-22), or rotary-wing (CH-53, CH-46, and UH-1) aircraft.

The Navy and Marine Corps also have amphibious craft that have demonstrated response capability in littoral areas, particularly those affected by flooding, earthquakes, and tsunamis.

MV-22 OSPREY	**C-130 HERCULES**
CH-53E SEA STALLION	**CH-46 SEA KNIGHT**
UH-1 IROQUOIS	

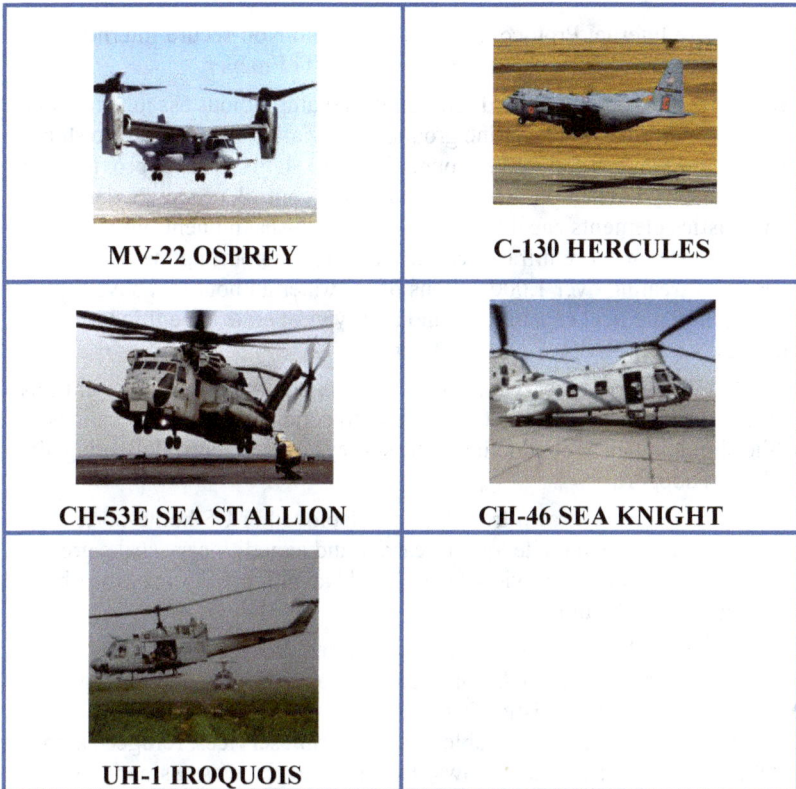

Figure 11-4: US Marine Corps Aircraft

11.6 Maritime Forces

Maritime forces can provide operational flexibility and assured access, while significantly reducing the footprint ashore and minimizing the permissions required to operate from the Affected State. Management of DOD maritime operations will be performed by the JTF Joint Forces Maritime Component Commander (JFMCC).

A Marine Expeditionary Unit (MEU) embarked aboard Amphibious Ready Groups (ARG), may be available to support FDR operations. These units provide a forward-deployed, flexible sea-based Marine Air Ground Task Force (MAGTF), capable of conducting: disaster relief, and crisis response. MEUs are characterized by their sea-based forward presence, expeditionary nature, and their interoperability with joint and combined forces. The MEU is ideally suited to provide immediate support for FDR operations. Every MEU deploys with a comprehensive communications suite providing voice,

11-9

data, Secure Internet Protocol Router (SIPR) and Non-secure Internet Protocol Router (NIPR) capability called a MEU Enabler.

MEUs typically deploy aboard three ships: an amphibious assault ship that holds the majority of the Marine ground and aviation forces and two ships that hold equipment and move ground forces to shore. This approximately 2,200-person MAGTF provides command and control, ground, aviation, and logistics elements capable of 15-days of self-sustainment, with minimal or no reliance on shore infrastructure. The ships supporting the MEU are capable of treating over 1,000 gallons of saltwater an hour and have engineering, medical capabilities, and ability to operate expeditionary airfields.

The MEU is able to quickly reposition or disaggregate into smaller units as the situation requires and has the capacity to conduct security operations while simultaneously conducting FDR surveys/assessments, delivering aid, and relief support.

Naval forces, can be drawn from forward deployed or US home-ported units. Those forces include aircraft carrier and expeditionary strike groups, surface action groups, individual ships, and Naval Expeditionary Combat Command (NECC) units.

NECC units include Naval Construction Forces, Maritime Security Squadrons, Maritime Civil Affairs Teams, Mobile Diving and Salvage units, Expeditionary Logistics Battalions, and Command and Control elements. Those units are capable of engineering services, refugee camp establishment, landward and seaward security, damage assessment, civil-military relations and CMOC operation, port openings, personnel recovery, air and surface ship cargo offloading and distribution, and planning, logistics, communication, and C2 staff-related functions.

Large deck amphibious ships are particularly useful for FDR operations as they can support both helicopters and small craft that can move supplies and equipment ashore. Those ships have organic medical support, C4I capabilities and berthing and messing facilities.

Military Sealift Command (MSC) ships provide sea-based support to forces afloat, including supply and hospital ships, as well as high-speed ferries and landing craft that can access areas beyond major ports. Specialized Joint Logistics Over the Shore (J-LOTS) equipment including causeways, crane ships and fuel systems can be of tremendous value to a heavily damaged port area, but have long lead times to establish.

The Navy also has expertise and equipment for port opening operations, including construction battalions, beachmasters, salvage and dive units and the Navy Expeditionary Logistics Support Group (NAVELSG). See Chapter 10.5.2.3 for more information on expeditionary port units.

NAVELSG is organized and staffed to provide a wide range of supply and transportation support critical for crisis response and humanitarian service missions. NAVELSG consists of a full-time support staff and five Navy Expeditionary Logistics Regiments and eleven Cargo Handling Battalions (NCHBs). NCHB is organized, trained and equipped to load and off-load Navy and Marine Corps cargo carried in maritime pre-positioning ships and merchant or container ships in all environments, operate an associated temporary ocean cargo terminal, load and off-load Navy and Marine Corps cargo carried in military controlled aircraft and operate an associated expeditionary air cargo terminal.

11.6.1 Maritime Airspace Control

Aircraft carriers and large amphibious ships are capable of providing maritime airspace control. Navy Helicopter Direction Control Centers are located on Landing Helicopter Assault/Landing Helicopter Dock (LHA/LHD)-class amphibious assault ships, see Figure 11-4. Those ships have robust communications capabilities and are capable of airspace command and control during FDR operations. They are able to serve as the JFACC or assist the appointed JFACC in air space planning, integration, and de-confliction.

Figure 11-5: Amphibious Assault Ship

A United States Marine Corps Air Support Squadron provides a Direct Air Support Center (DASC) cell for coordination and control of aircraft operating in direct support of a MAGTF. The entire DASC, or a portion of the cell, may deploy for FDR operations.

11.7 Engineering

FDR operations can be extremely engineer-intensive. In such cases, the JTF Commander may opt to establish a subordinate Task Force (TF) to control engineer operations and mission support tasks. The TF may be formed around an existing engineer command or naval construction regiment. The engineering assets attached to the subordinate TF will normally be constituted from all engineering resources (i.e., JTF and GCC Service components in the AOR).

The joint force engineer and staff may have to coordinate engineering activities with USAID/OFDA, the Affected State, NGO/IGOs, and appropriate cluster agencies. Some NGOs may have unique engineering capabilities that may be integrated into the overall operational effort.

NGO/IGOs may request extensive military engineer support for their activities and programs. It is critical to establish an effective engineer liaison in the Civil Military Operations Center (CMOC) to coordinate engineering support with these organizations.

The level of engineering assistance provided can vary from limited, highly specialized teams to complete engineer units. Limited teams are used to assess damage or estimate engineering repairs and can assist in specialized support such as power supply and distribution; utilities repair work, water purification, and well drilling operations. In large FDR operations, engineer units provide essential, general engineering support, including facility construction, structural repair, and camp construction for deployed forces.

Military engineers are frequently tasked to provide extensive cleanup and construction services and may be the only organization capable of providing that assistance. Requests for military engineering support should be coordinated, validated, and tasked through the USAID/OFDA MiTaM process. For additional information on engineer operations, refer to JP 3-34, Joint Engineer Operations.

11.7.1 Military Engineers

Military engineers provide engineering and related services in four broad areas.
- Military construction and support
- Engineering Research and development
- Water and natural resources management
- Support to other government agencies

In general, military engineering is separated into four distinct types of units:
- Combat engineers
- Civil engineers

- Horizontal engineers
- Vertical engineers

Each Service has an engineering organization oriented to its missions and personnel trained and equipped to fulfill those missions.

11.7.2 Engineering Assessments

The ability to analyze missions, both civilian and military, to produce the required outcome is a task required for engineer assessments.
USAID/OFDA uses those assessments to assist in mission validation and delegation of tasks.

11.7.3 Engineering Capabilities

11.7.3.1 Debris Clearing

Engineers, both combat and construction, have heavy equipment that is capable of removing debris, including:

- Snow and ice
- Mud and dirt
- Limbs and trees
- Vehicles
- Broken equipment
- Concrete
- Barrier material
- Waterborne debris in navigable waterways

11.7.3.2 Road, Ferry and Bridge Construction

If there is not sufficient time to contract construction work, or if contractors are not available, military engineers can build and repair roads and bridges for temporary use.

11.7.3.3 Fireline/Dozer Operations

Engineer units with bulldozers are capable of assisting with wildfire suppression using both equipment and manpower.

11.7.3.4 Shelter Construction

Military engineering construction units have the ability to construct temporary shelters with associated sewer, electrical, and water for internally displaced persons.

11.7.3.5 Power Generation

Engineers have power generation capabilities within most of their units. Mobile emergency power capability from 5 to 60 kilowatts or more is available.

> ⚠️ **WARNING:** *Military personnel are not permitted to connect military generators to civilian infrastructure. A certified civilian electrician must connect the power. Ensure that power lines are not reenergized by connecting infrastructure to generators.*

11.7.4 US Army Corps of Engineers

USACE provides an extensive range of expertise in:
- Engineering and construction support
- Critical infrastructure assessments
- Temporary repairs
- Temporary housing and critical public facilities
- Debris management

In fighting floods, USACE can:
- Assist in search and rescue operations
- Provide technical assistance and expertise
- Make emergency repairs to levees and other flood control projects
- May be able to provide flood fighting materials such as sandbags, plastic sheeting, lumber, pumps and rocks

Post-flood response activities of USACE may include:
- Clearing drainage channels, bridge openings, or structures blocked by storm-generated debris
- Clearing blockages to critical water supply intakes and sewer outfalls
- Clearing debris necessary to reopen vital transportation routes
- Restoring critical public services and facilities through temporary measures
- Identifying hazard mitigation opportunities

USACE can assist in rehabilitation by:
- Repairing and/or restoring completed levees, floodwalls, and other flood damage reduction projects
- Repairing and/or restoring hurricane or shore protection structures damaged or destroyed by wind, wave, or water action from storms

USACE provides assistance to military commanders through specially trained teams who rapidly respond to floods, hurricanes, earthquakes, or other disasters, whenever DOD has been tasked to assist in foreign disaster response around the globe.

Forward Engineering Support Teams (FEST) are the USACE elements that deploys to support engineer planning and mission execution. The FEST integrates USACE's capabilities with deployed engineer units to provide infrastructure planning, engineering design and assessment, contract construction, and environmental engineering to meet both military and civil requirements.

11.7.4.1 US Army Engineers

The military side of USACE is the largest engineering capability within DOD. The major engineering capabilities include keeping lines of communication and tactical march routes open, which may include continuous repair of damage caused by weather or heavy traffic.

Army Engineer Brigades capabilities include float-bridge assets for river-crossing operations.

Multi-Role Bridge Companies transport, assemble and operate ribbon rafts and float/fixed bridges during river-crossing operations.

Combat-Support Equipment Companies augment combat engineering units with equipment to move earth and maintain horizontal surfaces such as roads and airstrips. Combat engineers also assemble tactical bridges using panel –bridge companies or theater stocks.

Construction Engineers are capable of performing a variety of construction operations. The nature of FDR operations limits construction to essential facilities needed to sustain current operations. Missions include constructing:

- Forward logistics bases
- Heliports
- Main Supply routes
- Bed down sites

Specialized engineer units are capable of quarrying; electrical, water, gas, and sewer repair; water pipeline, petroleum pipeline and port construction.

Topographic engineering companies are capable of providing map products, and terrain-analysis for the JTF Commander and staff.

11.7.4.2 United States Army Geospatial Center

The United States Army Geospatial Center is part of USACE, and provides geospatial products such as maps, hydrological information, and virtual fly-over capabilities to provide units supporting FDR with up-to-date information on the disaster area.

11.7.5 Air Force Engineers

Air Force Combat engineers are capable of performing engineering-related aspects of air base operations. Basic missions of Air Force engineers include:

- Emergency repair (including rapid runway repair, facility repair, and utility repair)
- Bed down of Air Force units
- Operations and maintenance of Air Force facilities and installations
- Construction management
- Supply of materiel and equipment to perform the engineering mission

Air Force engineers are organized into three basic types of units: Red Horse, Prime Base Engineer Emergency Force (Prime BEEF), and Prime Readiness in Base Support (Prime RIBS). The engineering and services force module combines Prime BEEF and Prime RIBS capabilities to support a flying squadron.

Red Horse squadrons have a regional responsibility. They are mobile, rapidly deployable, and largely self-sufficient for limited periods of time. All *Prime BEEF* units are configured as squadrons and teams in order to provide peacetime facilities maintenance capabilities. Their mission is to provide support to the forces; these civil engineering units are organic at all major Continental United States (CONUS) and overseas Air Force bases. Prime BEEF units have no organic heavy equipment—only toolboxes and small team kits such as power tools. They require base operating support, and most deploy in 50 or 100-person teams.

Prime RIBS units are base support units and are not typically used in an FDR environment.

11.7.6 Navy Engineers

Naval Facilities Engineering Command (NAVFAC) delivers and maintains sustainable facilities; acquires and manages capabilities for the US Navy's expeditionary forces; provides contingency engineering response; and, enables energy security and environmental stewardship.

NAVFAC is supported by Atlantic and Pacific Commanders who also serve as Fleet Engineers. NAVFAC is comprised of ten Facilities Engineering Commands that have additional responsibilities as Regional Engineers.

Naval Construction Force (NCF) consists of a group of deployable naval units that has the capability to construct, maintain, and operate shore, inshore, and deep-ocean facilities in support of the USN and USMC, and

11-16

other agencies of the USG. The NCF is frequently referred to as the *Seabees*. Air-transportable, task-organized NCF units are capable of deployment with 48-hours notice. Additionally local contractual acquisition of heavy engineer equipment can augment air-transported equipment. NCF missions include:

- Military base construction support, including operational, logistics, underwater, ship-to-shore, shore facilities construction, maintenance, and operation
- Ship-to-shore construction support operations
- Damage repair operations
- Disaster control and recovery operations

11.7.7 US Marine Corps Combat Engineer Battalions

Each USMC division is supported by one Combat Engineer Battalion (CEB) that provides support to the division through task-organized engineer elements. The CEB can perform the following missions:

- Engineer reconnaissance
- Employing, repairing, and reinforcing bridge systems
- Construction support
- Repair and construction of roads
- Utility support, including mobile electric power equipment and potable water
- Constructing and improving vertical takeoff and landing sites

11.8 Communication

In an FDR environment, DOD missions may require use of multiple communication systems. Specific requirements may span the full capability of DOD communications abilities, including line of sight, over the horizon, and satellite communications systems for voice, video, and data. These assets, whether terrestrial or satellite based may be the only communications infrastructure in the area of operations thus heavily relied upon by the Affected State and the UN.

> **NOTE:** *USTRANSCOM's Global Standing Joint Force Headquarters (GSJFHQ) Joint Communications Support Element (JCSE) plays a key role in ensuring communications are established and maintained. The JCSE is capable of providing both NIPR and SIPR communications, and Government Emergency Telecommunications Service (GETS) (see Section 10.5.3 for additional information on the GSJFHQ. capabilities).*

Interoperability is the key to successful disaster relief operations. Non-mission capable or incompatible communications, overloaded command centers, distraught citizens, and exaggerated or inaccurate news media coverage contribute to confusion and chaos. See Chapter 5 for more information on coordination, collaboration, cooperation and communication.

Units responding to support civilian responders must be prepared to integrate communication systems with civilian agencies. Because of equipment differences, spectrum requirements, and the geography at the incident, commanders should not assume that tactical radio equipment is interoperable with civilian equipment.

Interoperability planning should include radio bridging devices that can connect varied devices such as tactical radios to cell phones, and sharing data through a common information management plan.

11.9 Civil Affairs

Civil Affairs (CA) units provide the JTF Commander with culturally-oriented, linguistically-capable personnel who can provide emergency coordination and administration capability when civilian political-economic structures within the Affected State have been incapacitated. These CA "generalists" possess a wealth of experience conducting a vast array of Civil-Military Operations (CMO) in permissive environments.

CA personnel are capable of supporting FDR operations in a variety of functional areas through planned CA Operations (CAO) or support to CMO. This support includes six categories of CA functional specialty areas:

1. Rule of law
2. Economic stability
3. Governance
4. Public health and welfare
5. Infrastructure
6. Public education and information

CA core tasks include:

- Support to civil administration
- Populace and resources control
- Disaster Relief
- Civil information management

CA assets can prove extremely valuable as advisors to the Commander on the impact of military activities within the civilian sector. CA personnel are trained to work with the local populace, support local disaster assistance activities, (i.e. contracting, establishing and maintaining liaison) with USG agencies and NGOs. CA assets assess infrastructure damage, assist in developing and managing temporary shelters, and coordinate activities with the CMOC. In the CMOC, CA personnel serve as liaison between military, diplomatic, and NGO participants.

Early deployment of CA assets can be a force multiplier, setting the stage for the introduction of follow-on forces.

CA functions should be represented on the JTF staff and its subordinate units to ensure CA is integrated into JTF plans and orders. CA planners can also assist in conducting assessments. CA personnel may also participate in a Joint Civil-Military Operations Task Force (JCMOTF). See JP 3-57, Civil-Military Operations, which provides specific guidance on planning and executing Civil Affairs operations.

11.9.1 Civil Affairs Service Commands

Conventional CA forces and their Service Commands are primarily composed of Reservists. CA specialists are trained to quickly identify critical requirements needed by local citizens in disaster situations. They also work with the Country Team and USAID representatives to locate resources and coordinate CMO. Although each CA Service Command is different, all provide JTF Commander with CA capabilities.

11.9.1.1 US Army Civil Affairs

The 95th CA Brigade, assigned to USSOCOM and stationed at Fort Bragg, NC is currently the Army's only Active duty CA Brigade (for more information on USSOCOM CA units see Section 10.3.2).

The majority of the Army's conventional forces CA personnel are in the Reserve Component (RC). The determination to mobilize RC CA assets must be a consideration during pre-deployment planning, because the request for forces and mobilization process for RC personnel is a lengthy process.

11.9.1.2 United States Army Civil Affairs and Psychological Operations Command

United States Army Civil Affairs and Psychological Operations Command (USACAPOC) is comprised of 12,000 Soldiers, and constitutes 94 percent of DOD's CA forces and 71 percent of DOD's MISO forces.

USACAPOC is located at Fort Bragg, NC, and is the headquarters for all Army Reserve CA and MISO units. USACAPOC units provide support to theater and task force commanders.

> **NOTE:** Currently, the Army has approximately 7600 Active and Reserve CA personnel. The Army is in the process of adding an additional 1,100 active-duty CA specialists to its ranks. By fiscal year 2013, the US Army CA community will consist of approximately 6400 Reserve, 1400 Active Duty, and 1500 Special Operations Forces personnel.

11.9.1.3 Maritime Civil Affairs and Security Training Command

Maritime Civil Affairs and Security Training (MCAST) Command trains, equips, and deploys naval personnel to facilitate and enable the Navy component or JTF Commander to establish and enhance relations between Affected State government, NGOs, IGOs, and the civilian populace. MCAST Command executes civilian-to-military operations and military-to-military training, as directed, in support of security cooperation and security assistance requirements.

Maritime Civil Affairs (MCA) operations blend maritime-specific functional specialties with established Army and Marine Corps Civil Affairs missions. To avoid mission duplication, MCA leverages Navy capabilities in port operations, as well as harbor and channel construction and maintenance. MCA consists of less than 200 AC and 150 RC personnel. MCAST Command's four core competencies are:

- Maritime Civil Affairs Plans and Operations
- Maritime Civil Affairs Assessment
- Maritime Functional Area Expertise
- Security Force Assistance Mobile Training Teams (SFA MTT)

MCA forces can operate from ships, thereby providing operational maneuver and assured access, while significantly reducing the footprint ashore.

11.9.1.4 US Marine Corps

The US Marine Corps (USMC) has approximately 200 AC/RC CA personnel capable of supporting a Joint or Marine Air Group Task Force (MAGTF) Commander with CA personnel. USMC CA personnel are trained and organized to facilitate planning, coordination, and execution of CMO.

11.9.1.5 US Air Force

The USAF does not currently have CA units.

11.10 Medical

Military medical capabilities may be requested to augment or sustain Affected State medical assets in order to save lives and minimize human suffering. This section gives a brief overview, by Service, of some of the capabilities available, and is not intended to be an all-inclusive list. See Chapter 7, Medical Officer, for additional information.

> *NOTE: DOD medical assets may not always be available and access to them may have to compete with on-going operational requirements. In addition, the deployment and set-up of medical resources is a lengthy process; it could take days or weeks before medical capabilities are fully established and ready to receive patients following a disaster.*

11.10.1 United States Army Medical Capabilities

11.10.1.1 Hospitalization

Army Combat Support Hospitals (CSH) are Medical Treatment Facilities (MTF) capable of providing hospitalization and outpatient services for 248 patients when fully staffed and equipped. Capabilities include emergency treatment, triage, and preparation of incoming patients for surgery; general, orthopedic, thoracic, urological, gynecological, and oral maxillofacial surgical capability; consultation services; pharmacy, psychiatry, community health nursing, physical therapy, clinical laboratory, blood banking, radiology, and nutrition care support; emergency dental treatment; medical administrative and logistical services; and laundry for patient linens.

11.10.1.2 Medical Evacuation (MEDEVAC) (Ground)

Ground ambulance companies are capable of providing ground evacuation support for litter and ambulatory patients. Capabilities also include movement of patients between CSH, aeromedical staging facilities, aeromedical staging squadrons, mobile aeromedical staging facilities, and railheads or seaports; emergency movement of medical personnel and supplies; and medical evacuation of wounded or injured personnel from the point of injury to supporting ambulance exchange points or MTFs.

11.10.1.3 Medical Evacuation (MEDEVAC) (Air)

Army air ambulance companies can provide air evacuation support throughout the Area of Operations (AO). Air ambulance companies have twelve to fifteen helicopter ambulances, and four or five forward support medical evacuation teams. Their capabilities include air crash rescue support; expeditious blood product delivery, biological, and medical supplies to meet critical requirements; rapid movement of medical personnel, supplies, and equipment to support mass casualty requirements or emergency movement of patients between hospitals, aeromedical staging facilities, hospital ships, seaports, and railheads.

11.10.1.4 Veterinary Services

The Army medical detachment veterinary service support provides dispersed veterinary Role 1 and Role 2 (medical and resuscitative surgical care); veterinary Role 3 (comprehensive canine medical and surgical care); evacuation and hospitalization support for military and contractor working dogs; endemic zoonotic and foreign disease epidemiological surveillance and control; animal holding facility and kennel inspections; commercial food source audits for DOD procurement; food safety, quality, and sanitation inspections; food defense vulnerability assessments; food and water risk assessments; food microbiological and chemical laboratory diagnostics for supported units; and foreign humanitarian assistance programs in support of all branches of the Service throughout the AO.

11.10.1.5 Medical Logistics

Army medical logistics companies provide direct support for Class VIII (medical) supplies, as well as, field and sustainment-level medical equipment maintenance and repair. Unit capabilities also include reception, classification, and issue of up to 11.1 short tons of Class VIII (medical); storage for up to 51 short tons of Class VIII (medical) supplies; building and positioning Class VIII (medical) support packages in support of contingencies; reconstitution of medical logistics units, sections, or teams; and coordination for emergency delivery of Class VIII (medical) supplies. When directed, the unit can also perform the single integrated medical logistics management supply and requisition processing mission for all joint forces in the theater.

11.10.1.6 Area Medical Support

Area support medical companies are MTFs capable of providing support for treatment of patients with disease and minor injuries, triage of mass casualties, initial resuscitation/stabilization, advanced trauma management, and preparation for further evacuation of ill or injured patients. Capabilities also include evacuation of patients; emergency medical supply and resupply; behavioral health consultation, to include operational stress control elements; pharmacy, laboratory and radiological services; emergency dental care to include stabilization of maxillofacial injuries and limited preventive dentistry; and patient holding capability.

11.10.1.7 Preventive Medicine

Preventive medicine detachments provide preventive medicine support and consultation for the prevention of disease; field sanitation, entomology, sanitary engineering; occupational and environmental health surveillance, and epidemiology to minimize the effects of environmental injuries, and enteric diseases, vector-borne disease, and other health threats with deployed forces. Capabilities also include monitoring pest management; field sanitation; water treatment and storage; waste disposal; other environmental health-related problems; recommending corrective measures; medical surveillance activities to assist in evaluating conditions affecting the health of the supported force; and, epidemiological investigations. The detachment also monitors: casualties, hospital admissions, and reports of autopsies for exposure to harmful agents (such as radiation, chemical, and biological hazards).

11.10.1.8 Combat and Operational Stress Control

Army Combat and Operational Stress Control detachments are capable of providing combat and operational stress control prevention and treatment services on an area basis. Capabilities include preventive consultation and stress education support; neuropsychiatric care, triage, and stabilization; assistance to non-medical units with operational stress reaction casualties; and critical events debriefing support. The detachments also have patient holding capability.

11.10.2 United States Navy Medical Capabilities

11.10.2.1 US Navy Amphibious Assault Ships

While the primary mission of amphibious assault ships is to support Marine landing forces, large-deck amphibious ships may be designated as Casualty Receiving and Treatment Ships (CRTS) and used for humanitarian and disaster relief missions. Primary CRTS (LHA/LHD) have laboratory (including blood) and radiology capability to support surgical suites. During disaster relief operations, primary CRTS are staffed to provide extensive medical and trauma support. Additional amphibious assault ships may be designated as secondary CRTS. These may include any class ship with the capability to receive and treat casualties, if appropriate medical materiel and personnel are available to provide care. Ships normally designated as secondary CRTS include Landing Platform Dock (LPD), Landing Ship Dock (LSD), and Load Carrying Capacity (LCC) class ships. The LHDs and the LHAs have the largest medical capability of any amphibious ship currently in use. They are capable of receiving casualties from helicopters and waterborne craft. Major medical facilities aboard those ships include four main operating rooms, two emergency operating rooms, and dental operating rooms; a 15-bed intensive care unit and a 45-bed patient ward; multiple dressing stations; and, a casualty collecting area on the flight deck. The ships also have medical elevators to transfer casualties from the flight deck to the ship's medical facilities.

11.10.2.2 US Navy Hospital Ships

The US Navy Mercy-class hospital ships are capable of providing a full-service deployable hospital asset for use by the military and other US government agencies involved in the support of disaster relief and humanitarian operations worldwide. US Naval Ships (USNS) Mercy (T-AH 19) and Comfort (T-AH 20) each contain 12 operating rooms and a 1,000 bed hospital facility (when staffed to that level); CAT-scan and digital radiological services; medical laboratory; pharmacy; 2 oxygen producing plants; a flight deck capable of landing military helicopters; and a side port to take on patients at sea. Limiting factors with their employment in FDR operations include the requirement to source staffing from major military treatment facilities (restricting or shutting down essential services) and the "tyranny of distance" (i.e., long deployment and transit times limit the utility of the platform for disaster affected populations).

11.10.2.3 US Navy Fleet Surgical Team

The mission of Fleet Surgical Teams (FST) is to provide surgical specialty augmentation to the primary CRTS platforms. They are distinct, free-standing assets of the Atlantic and Pacific operating forces. When not aboard ship, FST members remain in an additional duty status at regional Navy military treatment facilities. There are nine FSTs, with teams located in Norfolk, San Diego, and Okinawa. The standard composition of an FST includes two surgeons, an anesthesiologist, four physicians (Family Practice, Internal Medicine, Emergency Medicine, and Pediatrics), two nurses, a medical regulating officer, and nine enlisted support staff. FSTs are capable of being tailor-force packaged by specialty (e.g., orthopedic, pediatric, or trauma surgery) to meet specialized requirements of a disaster relief operation.

11.10.3 United States Air Force Medical Capabilities

11.10.3.1 US Air Force Small Portable Expeditionary Aeromedical Rapid Response Team

Small Portable Expeditionary Aeromedical Rapid Response (SPEARR) teams provide a rapid response, extremely mobile and highly clinically capable medical assets in support of a wide spectrum of contingency

11-24

missions. SPEARR Teams provide force health protection and direct medical support for an initial period of five to seven days to a population-at-risk (PAR) that may be comprised of all US military personnel or include a combination of international military and civilian personnel in a coalition operation. Sustainment or resupply capability ensures continued medical care and force health protection.

The scope of care includes public health and preventive medicine, flight medicine, primary care, emergency medicine and surgery, perioperative care, critical care stabilization, patient preparation for aeromedical transport, and aeromedical evacuation coordination/communication.

The SPEARR Team is capable of being ready for deployment within two hours of initial mission notification. The two-hour response time is dependent on the collocation of personnel and equipment and availability of standing "on call" team.

The team functions as an Emergency Medical Support (EMEDS) module, which is comprised of four teams: a Preventive and Aerospace Medicine (PAM) Advance Echelon (ADVON) Team, the Mobile Field Surgical Team, the Expeditionary Critical Care Team, and the equipment only Expanded Capability and Infrastructure Module. The team may deploy in a man portable mode (backpacks, medical bags, and personal equipment only) without the expanded capability and infrastructure module or in a one pallet equivalent trailer mode, which allows independent operations for five to seven days.

11.10.3.2 US Air Force Mobile Aeromedical Staging Facility

Mobile Aeromedical Staging Facility (MASF) are capable of providing rapid response patient staging, limited holding, and AE crew support capability to support humanitarian and civil disaster response operations.

The MASF is normally located at or near airheads capable of supporting mobility airlift. The MASF provides forward support with the smallest footprint. It consists of three short crews, communications, liaison, and patient care teams.

The MASF includes a capability to receive patients, provide supportive patient care, and meet administrative requirements on the ground while awaiting AE. Critical Care Air Transport Teams (CCATTs) are assigned to every forward based MASF to enhance rapid evacuation. The MASF brings tents for patient care but will often use a building of opportunity to conduct AE operations.

The communications capability assigned to the MASF can be integrated into the tanker/airlift control element or operations cell. An individual may be identified to work with the aerial port element on the flight line to coordinate AE load planning, configuration, and operational support.

MASF Capabilities include:

- Patient reception
- Medical care to patients transiting through the AE system
- Supportive nursing care
- Administrative support
- Self-supporting tasks
- Holding capability for patients entering the AE system

The MASF is equipped and staffed for routine processing of up to 50 patients at a time and can process a maximum of 140 patients every 24 hours. Because it has no beds, patients remain on the litters provided by the originating facility.

11.10.3.3 US Air Force Aeromedical Evacuation Liaison Team

Aeromedical Evacuation Liaison Team (AELT) provides support between the forward units and the AE system as the operational and clinical interface. AELT operational and clinical interface may occur at locations such as forward operating bases and aboard ship. The flight nurse on the team assists the medical unit in preparing AE patients for flight. The administrative officer is responsible for working with the airlift center and aerial port element to ensure the aircraft is properly configured, and equipment pallets, patients, and AE support personnel are properly manifested on the AE mission. Communications personnel will be integrated into the airlift operations element supporting flight line operations or the Wing Operations Center (WOC). Establishing a communication network with airlift operations is essential for rapid evacuation.

11.10.3.4 US Air Force Medical Global Reach Laydown Team

Medical Global Reach Laydown Team provides preventive medicine support to:

- Expeditionary Site Surveys – Tactical Air Coordination Center (TALCE) ADVON Team initial laydown activities
- Gas and Go – Routine TALCE aircraft refueling operations
- Bare Base – TALCE personnel perform pre-deployment site survey to support follow-on force build-up
- Air Bridge – A series of en-route locations outlining an air route of travel for rapid deployment and sustainment of forces

- Contingency Response Group (CRG) – Medical Global Reach Laydown Team is assigned to CRG to provide medical support during rapid opening of contingency airfields; purpose is to bring significant order, foresight, speed, and safety during the crucial opening days of a contingency.

The Medical Global Reach Laydown Team deploys with the TALCE/Medical Strike Team (MST) and assesses the health risks associated with environmental and occupational health hazards. More specifically, in support of establishing a potential Main Operating Base (MOB) in a forward deployed location, the Medical Global Reach Laydown Team:

- Provides medical input into the proper lay-down of installation facilities
- Determines adequacy of local billeting and public facilities
- Evaluates local medical capabilities
- Recommends locations for medical facilities
- Evaluates the safety and vulnerability of local food and water sources
- Assesses occupational and environmental hazards
- Performs vector/pest risk assessment
- Provides medical assessments
- Performs epidemiological risk assessments
- Performs health risk assessments
- Provides limited medical support, emergency planning and response

11.10.4 Patient Movement

To move casualties out of a disaster area, Affected State authorities will coordinate with supporting agencies to establish casualty collection points, from which patients will be moved to appropriate levels of care based on the severity of their injuries. Casualty transportation to support that movement may be either dedicated or lift-of-opportunity from any combination of government, commercial, and private resources. Affected State authorities will need to leverage available ground, rail, and both fixed and rotary-wing assets to support movement of disaster victims to the closest available MTFs.

DOD may provide support to evacuate seriously ill or injured patients away from the disaster, but within the Affected State. The vast majority of that support will use ground transportation or rotary-wing assets. On very rare occasions, DOD casualty movement and transportation assets (including ships and fixed-wing strategic-lift platforms) may be used to move patients to regional nations that are willing to accept victims, or to the US for definitive care.

USTRANSCOM may provide strategic lift for the patient movement mission, and the Global Patient Movement Requirements Center may deploy a Joint Patient Movement Team (JPMT) to the area. A JPMT regulates and tracks all patients transported on DOD assets to reception and medical treatment facilities. USTRANSCOM coordinates DOD transportation assets and establishes aeromedical evacuation centers at both APOE and APOD locations.

11.10.5 Mortuary Affairs and Mass Fatality Management

GCCs are responsible for giving authoritative direction and guidance to the JTF Commander with regard to mortuary affairs (MA) and for designating a Service component to serve as the lead for the Theater MA Support Program. Individual unit commanders sustaining losses are responsible for recovery and evacuation of human remains to the nearest designated MA collection point. Every effort must be made to identify and account for recovered human remains of US military personnel, government employees, government contractors, dependents, and other US citizens who die in foreign areas subsequent to natural disasters.

In certain cases where large scale, high-impact disasters cause very large numbers of fatalities in an Affected State, the GCCs may make mass fatality management (MFM) assistance available. That assistance could involve the provision of logistical resources such as human remains' pouches or cold storage facilities for setting-up temporary morgues, or engineering support to assist local authorities with mass internment measures when mass fatalities overwhelm Affected State mortuary support capacity.

Unless otherwise directed by the Secretary of Defense, it is DOD policy that only dedicated Service MA personnel handle human remains. The Secretary of Defense may approve the use of non-MA personnel to support MA personnel in certain mass fatality scenarios.

More detailed information may be found in *JP 4-06, Mortuary Affairs in Joint Operations*.

11.10.6 Department of Defense Mortuary Affairs Units

The DOD's MA capacity is very limited. The US Army has two Active and two Reserve Component Mortuary Affairs units stationed at Fort Lee, Virginia and Puerto Rico, respectively. DOD fixed-base Mortuary Affairs facilities include:

- Joint POW and MIA Accounting Command, Hawaii
- Armed Forces Medical Examiner's Office

- Armed Forces Institute of Pathology
- Mortuary Affairs Squadron (MA), Dover Air Force Base

Support provided by DOD MA personnel will be tailored to the needs of the requesting authority. All DOD FDR MA/MFM support will be conducted in close coordination with appropriate authorities designated as the Lead Agency for MA/MFM by the Affected State.

11.11 Force Protection

Force protection is provided by the Affected State and must be coordinated through the Embassy. In some cases, military forces may be requested to assist with security by the Affected State. Ground forces can also be instrumental for providing protection for the joint force and security for civilians, both victims and relief workers, as well as for OGAs, IGOs, and NGOs. Specific forces to assist in force protection such as military police maybe required if an environment changes from permissive to uncertain or hostile. Forces assigned a security mission will not take part in direct humanitarian assistance missions.

11.11.1 Security and Self Defense

Military forces have the inherent right of self-defense, and responsibility to protect themselves and military assets at all times. See Appendix A, the Legal Aspects of FDR for additional information.

> **NOTE:** *Use of Force is discussed in Appendix A, Standing Rules of Engagement. In accordance with CJCSI 3121.01B, SROE must be briefed to all military personnel prior to deployment. If directed to perform security tasks, and if trained and properly equipped, military personnel may use non-lethal weapons as a force protection option, subject to applicable ROE.*

11.11.2 Military Law Enforcement Units

On rare occasions, Military Law Enforcement (MLE) units may be tasked to support FDR operations in a force protection or security role. MLE officials are trained and equipped for decentralized operations and operate in highly mobile vehicles equipped with reliable communications systems.

NOTE: *All law enforcement actions should be performed by the Affected State. In some cases, the Affected State may request assistance with security. In such instances, the Country Team, DOS, and USAID/OFDA will decide upon appropriate support and provide coordination, as the situation dictates.*

11.11.2.1 US Army Military Police

The United States Army Military Police (MP) Corps is the largest of all the DOD MLE organizations. MPs provide expertise in law and order and stability operations in order to enhance security and enable mobility. MPs have five main functions:

- Maneuver and mobility support operations
- Area security operations
- Law and order operations
- Internment and resettlement operations
- Police information operations

Military Police Company's execute a wide variety of missions including:
- Mobility support for both vehicles and personnel
- Performing security for critical personnel, sites, cargoes, and railways
- Securing, safeguarding, sheltering and managing internally displaced persons
- Restoring order in civil disturbances
- Limited law and order operations
- Liaise, coordinating, and training for aspects of law enforcement to other agencies
- Disseminating information
- Providing force protection and security

11.11.2.2 US Navy– Shore Patrol/Masters at Arms

The United States Navy (USN) has two law enforcement elements the Shore Patrol (SP) and Master of Arms (MA). SP is an additional duty assigned to sailors to maintain order when a ship's crew is on liberty. MAs are a traditional military police. MAs perform antiterrorism, force protection, physical security, and law enforcement duties on land and at sea. Neither SP or MA typically support FDR operations.

11.11.2.3 United States Marine Corps Military Police

United States Marine Corps (USMC) military police and corrections personnel provide the commander continuous support in enforcing the law. Like their Army counterparts, USMC military police are frequently used as physical security resources in FDR operations. Missions include:

- Preventing and suppressing crime
- Assessing physical security posture
- Preserving military control
- Quelling disturbances
- Investigating offenses
- Apprehending offenders
- Protecting property and personnel
- Providing flight line security
- Investigating traffic accidents
- Controlling traffic
- Antiterrorism protection
- Handling and safeguarding IDPs, refugees or evacuees

11.11.2.4 United States Air Force Security Forces

United States Air Force (USAF) Security Forces are the military police and the air base ground defense forces of the USAF. They are responsible for ensuring the safety of all base weapons, property, and personnel. Security Forces personnel train dog teams and are occasionally assigned to an armory to safeguard arms, ammunition, and equipment. Security Forces Group provides a highly trained, rapidly deployable "first-in" force protection capability for operating locations in support of the USAF Global Engagement mission. Missions include:

- Nuclear and non-nuclear weapon system security
- Physical and information security
- Integrated base defense
- Law enforcement
- Antiterrorism, resource protection
- Corrections

UNCLASSIFIED

This page intentionally left blank

UNCLASSIFIED

APPENDIX A - LEGAL ASPECTS OF FOREIGN DISASTER RELIEF OPERATIONS

This appendix discusses the legal aspects of Foreign Disaster Relief operations and includes components of fiscal, international, and US Government (USG) laws, policy and presidential guidance.

A - 1 Statutory Authority

US agencies can only participate in foreign disaster relief missions under appropriate legal authorization, the references for which are listed below. A fundamental precept of fiscal law is that public funds may only be expended when authorized by Congress.

> **NOTE:** *The principal authority for DOD to conduct FDR is the Foreign Assistance Act of 1961. Other essential guidance includes Executive Order 12966 (Section 1.2.), Secretary of Defense Guidance (Section 1.5. 1.2) and Joint Doctrine contained in Joint Publication 3-29, Foreign Humanitarian Assistance.*

Foreign Assistance Act of 1961 (Public Law 87-195) provides the legal guidance for USG engagement with friendly nations. The FAA directs the DOS to provide policy guidance and supervision of programs created within the FAA. It also provides for certain members of the US Armed Forces to come under the Chief of Mission (COM) authority.

The Foreign Service Act of 1980 (Public Law 96-465) § 207 (22 USC. 3927) defines Chief of Mission authority over executive branch personnel in their countries, as well as the basic relationships between the Department of State and other departments, agencies, and offices of the US Government.

The Diplomatic Security Act of 1986 (Public Law 99-399) charges the Secretary for the security of US Government operations and personnel abroad, empowering the Secretary of State to coordinate US Government personnel and establish appropriate staffing levels for missions

The Foreign Affairs Reform and Restructuring Act of 1998 (Public Law 105-277) provided for the reorganization of Arms Control & Disarmament Agency (ACDA) and United States Information Agency (USIA) functions within the Department of State and certain other actions concerning USAID

A - 2 Presidential Directives

Executive Order 12966, Foreign Disaster Assistance governs the implementation of § 404 of Title 10, United States Code, as amended. It also directs the Secretary of Defense to respond to man-made or natural disasters when the Secretary of Defense determines that such assistance is necessary to prevent loss of lives. It also authorizes the Secretary of Defense to provide disaster assistance only:

- At the direction of the President
- With the concurrence of the Secretary of State, or
- In emergency situations in order to save human lives, where there is not sufficient time to seek the prior initial concurrence of the Secretary of State, in which case the Secretary of Defense shall advise, and seek concurrence of, the Secretary of State as soon as practicable thereafter

National Security Presidential Directive 44 – Management of Interagency Efforts Concerning Reconstruction and Stabilization promotes the security of the United States through improved coordination, planning, and implementation for reconstruction and stabilization assistance for foreign states and regions at risk of, in, or in transition from conflict or civil strife. It also encourages agencies to coordinate and strengthen efforts of the United States Government to prepare, plan for, and conduct reconstruction and stabilization assistance and related activities in a range of situations that require the response capabilities of multiple United States Government entities to harmonize such efforts with US military plans and operations. To this end, the Secretary of State leads integrated US government efforts. However, this does not affect the authority of the Secretary of Defense or the command relationships of the US armed forces. This NSPD is important because it replaced Presidential Decision Directive 56, which established a precedent requiring USG interagency coordination, cooperation, and support.

National Security Decision Directive (NSDD) 38 – Staffing at Diplomatic Missions and Their Overseas Constituent Posts provides the Chief of Mission with the authority to determine the size, composition, or mandate of personnel operating under their authority.

The President's Letter of Instruction to Chiefs of Missions is sent to each COM upon assumption of office and charges the Chief of Mission to exercise responsibility for executive branch personnel in his or her country and to protect all US Government personnel on official duty abroad.

A - 3 Status of Forces Agreements

The United States has been party to multilateral and bilateral agreements addressing the status of US armed forces while present in a foreign country. These agreements are commonly referred to as Status of Forces Agreements (SOFAs). A SOFA is an agreement that establishes the framework under which armed forces operate within a foreign country. The agreement provides for rights and privileges of covered individuals while in the foreign jurisdiction, addressing how the domestic laws of the foreign jurisdiction shall be applied to US personnel while in that country. It is important to note that a SOFA is a contract between parties and may be cancelled at the will of either party. SOFAs are peacetime documents and therefore do not address the rules of war, the Laws of Armed Conflict, or the Laws of the Sea. In the event of armed conflict between parties to a SOFA, the terms of the agreement would no longer be applicable.

SOFAs may include many provisions, but the most common issue addressed is which country may exercise criminal jurisdiction over US personnel. The United States has concluded agreements where it maintains exclusive jurisdiction over its personnel, but more often the agreement calls for shared jurisdiction with the receiving country. In general, a SOFA does not authorize specific exercises, activities, or missions. Rather, it provides the framework for legal protections and rights while US personnel are present in a country for agreed upon purposes. A SOFA is not a mutual defense agreement or a security agreement. The existence of a SOFA does not affect or diminish the parties inherent right of self-defense under the law of war.

DOD medical personnel who have a current, valid, and unrestricted license to practice medicine, osteopathic medicine, dentistry, or another health profession and who are properly licensed under Title 10 USC § 1094(d) may practice their profession on non-DOD personnel at any location authorized by the Secretary of Defense.

Many existing Status of Forces Agreements have a statutory provision that addresses (1) the recognition of medical licenses issued by another country, or (2) the waiver of the country's licensure requirements for DOD medical personnel who enter the state solely to provide medical treatment to civilian victims of an emergency or disaster incident.

The United States is currently party to more than 100 agreements that may be considered SOFAs. A SOFA as a stand-alone document may not exist with a particular country, but that does not necessarily mean that the status of US personnel in that country has not been addressed. Terms commonly found in SOFAs may be contained in other agreements with a partner country so that a separate SOFA is not always utilized.

A - 4 Important Memoranda

Department of State - Department of Defense Memorandum of Understanding of 1997

Department of State - Department of Defense Memorandum of Understanding of 1997 revises an earlier 1996 MOU and covers the general security of certain DOD elements and personnel not under the command of a geographic combatant commander.

Department of State - Justice - Treasury Memorandum of Understanding of 1996

Department of State - Justice - Treasury Memorandum of Understanding of 1996 sets forth the authorities of the Chief of Mission (COM) in relation to law enforcement personnel abroad and outlines agreed principles with respect to the coordination of law enforcement.

A - 5 Fiscal Law

Resourcing and Reimbursement- the Defense Security Cooperation Agency/Office of Humanitarian, Disaster, and Civic Aid Funding

As discussed in Chapter 10 within the Defense Security Cooperation Agency (DSCA) section, the Office of Humanitarian Assistance, Disaster Relief, and Mine Action provides supervision and oversight of DOD humanitarian mine action and humanitarian assistance programs for the Director for Programs, DSCA.

Humanitarian programs support foreign policy and national security interests of assuring our allies and friendly nations and dissuading would-be aggressors by enhancing the legitimacy of the Affected State, by improving its capacity to provide essential services, and regional security and stability. DOD humanitarian assistance programs provide Combatant Commanders with unobtrusive, low cost, but highly effective, instruments to carry out their security cooperation missions. DSCA's role in FDR is assisting in capturing cost, reimbursement, and resourcing GCCs for the mission.

Title 10 § 404, Overseas Humanitarian, Disaster and Civic Aid (OHDACA) funding is the means by which DSCA reimburses Geographic Combatant Commands and the Services for United States Agency for International Development/Office of US Foreign Disaster Assistance (USAID/OFDA) and command-validated FDR mission support. DSCA reimburses incremental costs that would not have been incurred had the FDR operation not been supported.

To ensure FDR missions do not negatively affect DOD operational and maintenance budgets, it is critical that accurate accounting of expenses takes place; this will facilitate timely DOD reimbursements from the Department of State/USAID.

Bills and vouchers shall be processed by the Military Departments and forwarded as requested by the DOD Coordinator for Foreign Disaster Relief who will arrange to have them aggregated and forwarded to the Department of State for payment. *DODD 5100.46, December 4, 1975.*

When preparing billings for reimbursement of costs incurred, the following guidelines apply:

- Materials, supplies, and equipment from stock will be priced at standard prices used for issue to DOD activities, plus applicable accessorial costs for packing, crating, handling, and transportation
- Materials, supplies and equipment determined to be excess to the DOD will be available for transfer to the Department of State without reimbursement, in accordance with established DOD policies. Accessorial charges for packing, crating, handling, and transportation will be added where applicable
- Air and ocean transportation services performed by the Air Mobility Command (AMC) and the Military Sealift Command (MSC) will be priced, where applicable, at current tariff rates for DOD Components. Where tariff rates are not applicable, air transportation, whether provided by AMC or other aircraft, will be priced at the "Common-User Flying-Hour," rate for each type of aircraft involved and ocean transportation provided by MSC will be priced at "Ship Per Diem Rates"
- Services furnished by activities under DOD Industrial Funds other than AMC and MSC will be priced to recover direct and indirect costs applicable to reimbursements for services rendered to other Department of Defense activities
- Personal services furnished will be priced at rates to recover overtime of civilian personnel

All other services furnished, not specifically covered above, shall be priced on a mutually agreeable basis and, if feasible, such prices shall be established prior to the services being furnished. Prices for such services shall be at the same rates that the Department of Defense would charge other Government Agencies for similar or like services, if such rates are available; otherwise the basis of pricing will be to effect full reimbursement to the Department of Defense appropriations for "out-of-pocket" expenses.

A - 6 Other Potential Funding Sources

The following laws and statutes address other potential DOD funding sources. COCOM SJAs and Comptrollers are responsible for identifying those sources and securing proper authorities to utilize them:

- Emergency & Extraordinary Expense (10 USC § 127)
- Acquisition and Cross Servicing Agreements (10 USC § 2341, § 2342, § 2344)
- COCOM Initiative Fund (10 USC § 166(a))
- Transport of Humanitarian Supplies (10 USC § 402 Denton/Space-A) & § 2561(a)(2))
- Bilateral Cooperation Program (10 USC § 1050 & § 1051)
- Non Lethal Supplies for Humanitarian Relief (10 USC § 2547, § 2557, § 2562)
- Minor, Emergency and Contingency Construction (10 USC § 2803, § 2804, § 2805)
- Military Supplies and Services (10 USC § 7227, § 9626)
- Annual National Defense Authorization Acts

Fiscal law for FDR training and exercises (not operations) includes Humanitarian and Civic Assistance in Conjunction with Military Operations (10 USC § 401).

A - 7 Rules of Engagement

Rules of engagement (ROE) are directives issued by a competent military authority that delineate the circumstances and limitations under which US naval, ground, and air forces will initiate and/or continue combat engagement with other forces encountered (JP 1-02, *Department of Defense Dictionary of Military and Associated Terms*).

ROE provide a framework encompassing national policy goals, mission requirements, and the rule of law should the use of force become necessary. ROE may also regulate a commander's capability to influence a situation by granting or withholding the authority to use particular weapons, systems, or tactics.

- ROE do not direct the use of force; they authorize it
- When, where, and how to use force is the Commander's decision
- Well developed ROE provide maximum flexibility within the Commander's intent and guidance
- Flexible ROE stay ahead of the mission and enable success

A-6

- ROE development and management require a formal staff process
- ROE should be developed by operators and planners and supported by the staff judge advocate (SJA)
- Bi-lateral, multinational or coalition ROE require understanding treaty obligations, national policies, international law, cultural awareness, negotiation, consensus, and unity of effort

At a minimum, in permissive FDR environments, ROE should:

- Enhance mission accomplishment by supporting force protection while not hampering the agility of Department of Defense assets
- Emphasize restraint with clearly delineated escalation of force procedures
- Be professionally formulated, explained, and managed by formal staff practices

> **NOTE:** At the strategic level, separate ROE may be issued by the Joint Staff as part of the Chairman of the Joint Chiefs of Staff execution order. At the tactical and operational levels, ROE are developed and managed by the J3 and J5, both of whom are supported by the staff judge advocate.

For further reading on ROE, consult the UNCLASSIFIED portions of Chairman of the Joint Chiefs of Staff Instruction 3121.01B, 13 June 2005.

> **The Strategic Corporal:**
>
> *"A single act could cause significant military and political consequences; therefore, judicious use of force is necessary. Restraint requires the careful and disciplined balancing of the need for security, the conduct of military operations, and the national strategic end state."*
>
> *JP 3-0, Joint Operations*

A - 8 Legal References
- *Operational Law Handbook*. Charlottesville: US Army, 2010.
- CJCSI 3121.01B, "Standing Rules of Engagement for US Forces," 13 Jun 05
- Joint Pub 1-04 "Legal Support to Military Operations," 1 Mar 07
- MC 362, "NATO Rules of Engagement," 9 Nov 99

This page intentionally left blank

APPENDIX B - DEPARTMENT OF DEFENSE GUIDELINES FOR INTERACTION WITH NON-GOVERNMENTAL ORGANIZATIONS

The following Department of Defense (DOD) guidelines for interaction with Non-Governmental Organizations (NGO) are taken from best practices and lessons learned from military personnel engaged in Foreign Disaster Relief (FDR) operations, Service and Joint doctrine, as well as NGO sources.

Relationships between DOD and NGO personnel are based on multiple factors:

- a mutual understanding of each others' goals and objectives
- planned lines of operations
- support relationships
- coordination procedures
- information sharing
- capabilities (such as, dependent or independent; mature or immature)

Most NGOs follow the humanitarian principles of impartiality, independence, humanitarianism, and neutrality when providing aid; therefore, they base decisions on need alone. While not denying the necessity for occasional military assistance, some NGOs may perceive military aid as politically motivated and conditional. NGOs may be concerned that their association with civil-military or military organizations may undermine their impartiality and neutrality or increase their security risks. The collective aim should be to strive for *unity of effort* or at least *unity of purpose* through a comprehensive approach to achieving common objectives.

In spite of possible philosophical differences between military forces and NGOs, their short-term objectives are frequently similar. Discovering this common ground is essential. Attention to shared goals will facilitate cooperation, collaboration, coordination, and communication among DOD personnel and their partners within the NGO community.

B - 1 Cooperation and Coordination

Figure B-1 provides four possible scenarios for working with NGO personnel.

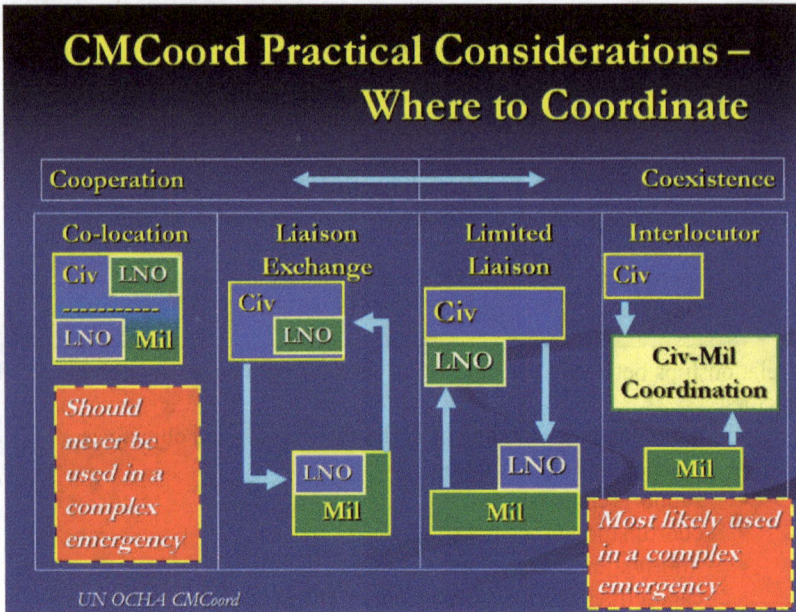

Figure B-1: Cooperation and Coordination between Military and Civilian Humanitarian Partners

Each scenario is situation-specific and should be based upon the following considerations:

- FDR objectives and available resources
- How the arrangement may affect the security of civilian humanitarian staff and beneficiaries
- Consideration of perceptions, accountability, and transparency

Scenario one entails humanitarian agencies and military units operating from within the same compound. While close proximity between partners can facilitate greater opportunities for information sharing and building relationships, co-location can lead to the perception that NGOs are affiliated with military personnel. This perception can have adverse security implications for the civilian humanitarian agency staff and beneficiaries. This scenario should only be used when the United States government is positively perceived by an Affected State.

Scenarios two and three involve liaison officers. In liaison exchange scenarios, liaison officers are assigned and work in the offices of another agency, and report back to their own agency. In limited exchange scenarios, liaison officers maintain an office in their own agency but travel to the other agency's office to conduct business. Both scenarios allow for team-building while providing greater transparency and an image that NGO partners are not affiliated with the military.

Scenario four provides the greatest NGO autonomy. In this instance, liaison officers maintain an office in their own agency and travel to a neutral site to conduct business, such as the United Nations (UN) Office for the Coordination of Humanitarian Affairs (OCHA) or a multi-purpose governmental ministry or neutral building not affiliated with either entity.

> **NOTE:** *Whenever possible, security permitting, military personnel should coordinate with humanitarian organizations in the most open forum possible, i.e. not on military compounds.*

B - 2 Humanitarian Operations Center

The HOC, as defined in JP 3-57, refers to a senior level international and interagency coordinating body that seeks to achieve unity of effort among all participants in large FDR operations. HOCs are horizontally structured organizations with no command and control authority, and all members are ultimately responsible to their own organizations or countries. The HOC has largely been replaced through the establishment of the UN Cluster System.

B - 3 Humanitarian Assistance Coordination Center

During FDR activities, the combatant command's crisis action organization may organize as a HACC to assist with the interagency and NGO coordination and planning. Staffing includes a director, a CMO planner, a USAID/OFDA advisor or liaison if available, a public affairs officer (PAO), contracting officer, engineer, NGO advisor, and other augmentation (e.g., legal advisor, surgeon, POL-MIL advisor) when required. Normally, the HACC is temporary; once a CMOC or HOC has been established, the role of the HACC diminishes, and its functions are accomplished through the normal organization of the combatant command's staff and crisis action organization.

B - 4 Civil-Military Operations Center

The CMOC is a component of Civil Affairs units. CMOCs enable coordination at the tactical and operational level between representatives of US military forces, other governmental agencies, the private sector, and NGO partners. The CMOC does not set policy or direct operations. The organization of the CMOC is theater-specific and mission dependent; it is

B-3

flexible in size and composition. A commander at any echelon may establish a CMOC. For further information on CMOC composition, tasks, and capabilities, see JP 3-57, *Civil-Military Operations*.

Figure B-2 provides a notional composition of a CMOC. Figure B-3 provides comparisons between these entities.

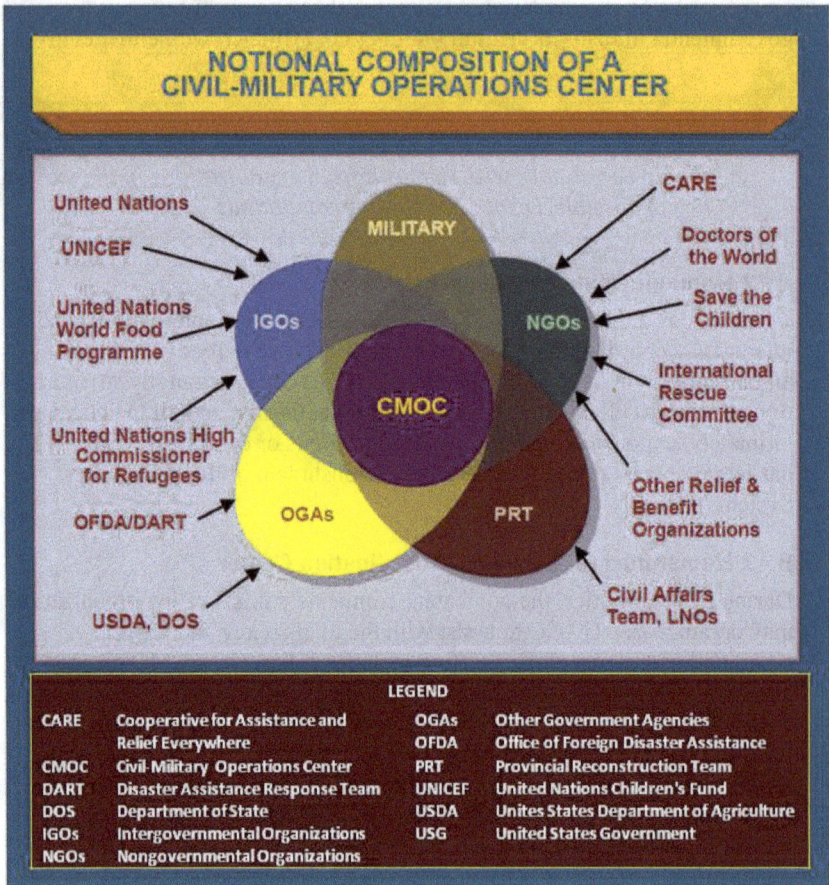

NOTIONAL COMPOSITION OF A CIVIL-MILITARY OPERATIONS CENTER

United Nations
UNICEF
United Nations World Food Programme
United Nations High Commissioner for Refugees
OFDA/DART
USDA, DOS

MILITARY
IGOs
CMOC
OGAs
NGOs
PRT

CARE
Doctors of the World
Save the Children
International Rescue Committee
Other Relief & Benefit Organizations
Civil Affairs Team, LNOs

LEGEND			
CARE	Cooperative for Assistance and Relief Everywhere	OGAs	Other Government Agencies
		OFDA	Office of Foreign Disaster Assistance
CMOC	Civil-Military Operations Center	PRT	Provincial Reconstruction Team
DART	Disaster Assistance Response Team	UNICEF	United Nations Children's Fund
DOS	Department of State	USDA	Unites States Department of Agriculture
IGOs	Intergovernmental Organizations	USG	United States Government
NGOs	Nongovernmental Organizations		

Figure B-2: Notional Composition of a CMOC

COMPARISON BETWEEN HUMANITARIAN OPERATIONS CENTER, HUMANITARIAN ASSISTANCE COORDINATION CENTER, AND CIVIL-MILITARY OPERATIONS CENTER

	ESTABLISHING AUTHORITY	FUNCTION	COMPOSITION	AUTHORITY
HOC	Designated Individual of Affected State, United Nations, of US Government Agency	Coordinates Overall Relief Strategy at the National (County) Level	Representatives from: Affected State; United Nations; US Embassy or Consulate; Joint Task Force; Other nonmilitary agencies, concerned parties (private sector)	Coordination
HACC	Combatant Commander	Assists with Interagency Coordination and Planning at the Strategic Level. Normally is disestablished once a HOC or CMOC is established.	Representatives from: Combatant Command; Nongovernmental Organizations; Intergovernmental Organizations; Regional Organizations; Concerned Parties (private sector)	Coordination
CMOC	Joint Task Force or Component Commander	Assists in Coordination of Activities at the Operational level with Military Forces, US Government Agencies, Nongovernmental Organizations, and Regional Organizations.	Representatives from: Joint Task Force; Nongovernmental Organizations; Intergovernmental Organizations; Regional Organizations; US Government Agencies; Local Government (Affected State); Multinational Forces; Concerned Parties (private sector)	Coordination

LEGEND
HOC - Humanitarian Operations Center
HACC - Humanitarian Assistance Coordination Center
CMOC - Civil-Military Coordination Center

Figure B-3: Comparison of the Structure and Function of the HOC, HACC, and CMOC

B - 5 Communication

Communication pertains not only to how information is shared between military and NGO personnel. It also includes guidelines for communicating the differences between DOD and NGO FDR activities to the affected population. See the Public Affairs Officer section of Chapter 6 for more detailed information on strategic messages.

B - 6 Information Sharing

The use of "open" communications and UNCLASSIFIED information-sharing networks are important for effective FDR operations. Because many DOD partners do not have a security clearance, it is crucial that whenever possible, information remain UNCLASSIFIED and available for unlimited distribution. Additionally, NGOs operating in a FDR environment may have minimal equipment for transmitting information. The quality and condition of an Affected State's communication network can further hamper information sharing efforts. Connectivity and network availability permitting, cellular phones and open-source platforms like SharePoint can be helpful in organizing portals and web pages for storing and sharing information between DOD and NGO partners.

The sharing of information in both directions is an essential element of successful FDR operations. Just as the military can provide logistical, transportation, and security information, NGOs can be an important source of information regarding local cultural practices that will bear on the relationship of military forces to the affected population.

Humanitarian organizations must maintain their neutrality and independence. Accordingly, some may not want to be seen with military forces. In such cases, the only coordination they may perform with the military is remote information sharing. Furthermore, because most NGOs have very little (if any) protective equipment for personal security, any activity that makes an NGO appear to be part of military operations can result in their becoming targets by hostile elements of a population.

B - 7 Communicating the Differences between DOD and NGO Personnel to the Local Population

Due to the level of interaction between DOD and NGO personnel in FDR operations, it can become increasingly difficult for an affected population to make a distinction between military and NGO activities. The following InterAction/DOD guidelines communicate these differences. Military personnel should abide by the following rules when working with NGO partners:

- Military uniforms should be worn
- NGO logos should not be worn or displayed on their person or equipment
- Visits to NGO sites should be by prior arrangement
- Security permitting, no weapons should be carried onto NGO sites
- Respect NGO views about meeting at military sites

- Never refer to NGOs as "force multipliers" or other kinds of military assets
- Do not interfere with NGO relief efforts, especially to "unfriendly populations"
- Respect NGO policies about being implementing partners for the military

NGO personnel should abide by the following guidelines when working with military units involved in FDR projects:

- Military uniforms should not be worn
- Visits to military sites should be by prior arrangement
- Travel in military vehicles should be limited to liaison personnel
- Should not be co-located with the military
- Should use their own logos when security conditions permit
- Should minimize their activities at military sites
- Requesting military support should be a last resort and provision of such support is not obligatory by the military

This page intentionally left blank

UNCLASSIFIED

APPENDIX C - REPORT FORMATS

C - 1 MiTaM

Request for Assistance. The Mission Tasking Matrix (MiTaM) process begins with an Affected State request for assistance to the United States Agency for International Development/Office of United States Foreign Disaster Assistance (USAID/OFDA). Should DOD possess a unique capability and assets are available to fulfill this need, then USAID/OFDA representatives will request DOD support. The USAID/OFDA representatives in coordination with the J-9 will develop the MiTaM request.

JTF MiTaM Processing. The MiTaM is processed by the JTF in five phases.

Phase 1 -MiTaM Receipt. The Joint Task Force (JTF), typically the Battle Watch Captain and Orders Officer, receives the MiTaM from USAID/OFDA and staffs it with the MiTaM Working Group (WG). Then, the MiTaM WG provides further analysis of the feasibility of the JTF's to conduct the mission.

Phase 2 - Staffing. The MiTaM WG is comprised of key JTF planners, unit representatives, and led by the J-9 in coordination with the J-3 (the J-3 is the tasking authority). The MiTaM WG validates the MiTaM, identifies units and resources capable of performing the mission, and issues a warning order. The Battle Watch Captain develops a tracking mechanism for MiTaMs and tracks them using their USAID/OFDA assigned number.

Phase 3 - Orders Development. The Orders Officer writes an Execution Order, identifying a list of corresponding tasks, including the MiTaM number for tracking; the Execution Order is issued to the tasked unit.

Phase 4 - Mission Execution. The tasked unit executes the assigned mission and the Battle Watch Captain tracks MiTaM progress.

Phase 5 – Closure. This phase includes capturing resources requirements, i.e., personnel, equipment operational costs, and consumables associated with mission execution.

> **NOTE:** *Tracking of MiTaM tasks should be conducted by J-3 Current Operations.*

Figure C-1 contains an example of a MiTaM.

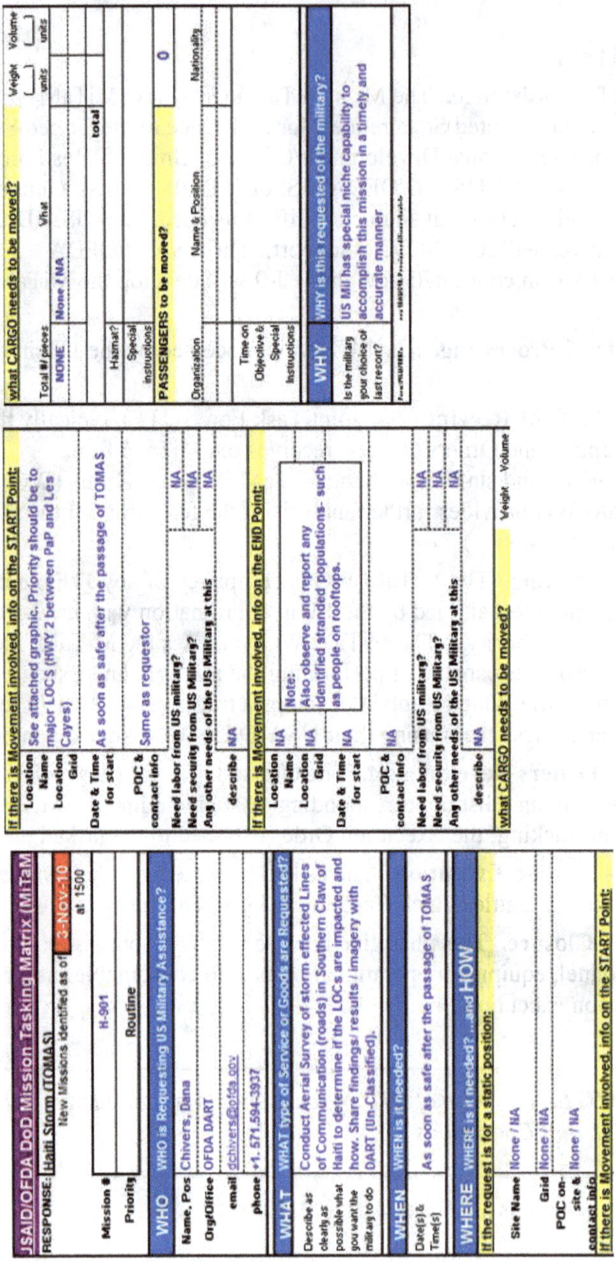

USAID/OFDA DoD Mission Tasking Matrix (MiTaM)

RESPONSE: Haiti Storm (TOMAS)

New Missions identified as of 3-NOV-10 at 1500

Mission #: H-901

Priority: Routine

WHO WHO is Requesting US Military Assistance?

Name, Pos: Chivers, Dana
Org/Office: OFDA DART
email: dchivers@ofda.gov
phone: +1.571.594.3937

WHAT WHAT type of Service or Goods are Requested?

Describe as clearly as possible what you want the military to do: Conduct Aerial Survey of storm effected Lines of Communication (roads) in Southern Claw of Haiti to determine if the LOCs are impacted and how. Share findings/ results / imagery with DART (Un-Classified).

WHEN WHEN is it needed?

Date(s) & Time(s): As soon as safe after the passage of TOMAS

WHERE WHERE is it needed? ...and **HOW**

If the request is for a static position:

Site Name: None / NA
Grid: None / NA
POC on-site & contact info: None / NA

If there is Movement involved, info on the START Point:

If there is Movement involved, info on the START Point:

Location: See attached graphic. Priority should be to
Name: major LOCS (HWY 2 between PaP and Les
Location: Cayes)
Grid:

Date & Time for start: As soon as safe after the passage of TOMAS

POC & contact info: Same as requestor

Need labor from US military? NA
Need security from US Military? NA
Any other needs of the US Military at this NA

describe: NA

If there is Movement involved, info on the END Point:

Location: NA
Name: NA
Location: NA
Grid:

Date & Time for start: NA

POC & contact info: NA

Need labor from US military? NA
Need security from US Military? NA
Any other needs of the US Military at this NA

describe: NA

Note:
Also observe and report any identified stranded populations- such as people on rooftops.

what CARGO needs to be moved? Weight Volume

What CARGO needs to be moved? Weight Volume units units

Total #/pieces What
NONE None / NA total

Hazmat?
Special Instructions

PASSENGERS to be moved? 0

Organization Name & Position Nationality

Time on Objective &
Special Instructions

WHY WHY is this requested of the military?

Is the military your choice of last resort? US Mil has special niche capability to accomplish this mission in a timely and accurate manner

Figure C-1: Example MiTaM

UNCLASSIFIED

C - 2 Personnel Status (PERSTAT) Worksheet

Figure C-2 contains an example PERSTAT worksheet.

JPERSTAT
AS OF TIME/DATE
MILITARY AND CIVILIAN PERSONNEL

COMMAND	USA	USAR	ARNG	USAF	USAFR	ANG	USN	USNR	USMC	USMCR	USCG	USCGR	MIL TOTAL	DOD CIV	DOD CONT	CIV OTHER	CIV CONT	CIV TOTAL	COMBINED TOTAL	PREV TOTAL	DELTA
GROUND FORCES																					
GROUND FORCES TOTAL	0	0	0	0	0	0	0	0	0	0	0	0	0	0	0	0	0	0	0	0	0
AFLOAT																					
AFLOAT TOTAL	0	0	0	0	0	0	0	0	0	0	0	0	0	0	0	0	0	0	0	0	0
COMBINED TOTAL	0	0	0	0	0	0	0	0	0	0	0	0	0	0	0	0	0	0	0	0	0
CASUALTY DATA																					
KIA	0	0	0	0	0	0	0	0	0	0	0	0	0	0	0	0	0	0	0	0	0
WIA	0	0	0	0	0	0	0	0	0	0	0	0	0	0	0	0	0	0	0	0	0
MIA	0	0	0	0	0	0	0	0	0	0	0	0	0	0	0	0	0	0	0	0	0
CAPTURED	0	0	0	0	0	0	0	0	0	0	0	0	0								
DUSTWUN (MIL)	0	0					0		0		0		0						0	0	0
EAWUN (CIV)	0	0					0		0		0		0						0	0	0
NON-HOSTILE DEATH	0	0					0		0		0		0						0	0	0
NON-HOSTILE INJURY	0	0	0	0			0	0	0	0	0	0	0	0	0	0	0	0	0	0	0
TOTALS	0	0	0	0	0	0	0	0	0	0	0	0	0	0	0	0	0	0	0	0	0
*** PLEASE NOTE: Ships or units that have departed (or removed from the count) will be highlighted in pink, the number of personnel moved out and subsequently removed from the JPERSTAT on the next day in order to track any increase or decrease in Delta.																					
RTD WIN 72 HOURS																					
REMARKS:																					

Figure C-2: Example PERSTAT

C - 3 Logistics Status (LOGSTAT) Worksheet

Figure C-3 contains an example LOGSTAT worksheet.

UNIT NAME — **As of:**

CLASS I

Type	UBL (DOS/Units)	O/H Balance or Current Rate	Status	Projected Requirement Next 24 hours	Projected Requirement Next 48 hours	Projected Requirement Next 72 hours
UGR(DOS)	3	3	🟢 100.0			
MRE(DOS)	3	2	🔴 66.7			
ICE(LBS)	500	350	🟡 70.0			
WATER(GALS)	1800	1500	🟡 83.3			
TWPS(GALS/HR)	100	60	🔴 60.0			

CLASS III (BULK)

TYPE	UBL (Gallons)	O/H Quantity (Gallons)	Status	Projected Requirement Next 24 hours	Projected Requirement Next 48 hours	Projected Requirement Next 72 hours
JP8	9000	9000	🟢 100.0			
MOGAS	2000	1000	🟡 50.0			
DIESEL	1000	750	🔴 75.0			

CLASS III (PACKAGE)

TYPE	NIIN	UI	Quantity On Hand	Projected Requirement Next 24 hours	Projected Requirement Next 48 hours	Projected Requirement Next 72 hours
CHLORINATION KIT FED	6850-00-270-6225	KT	50	50	40	30
ANTIFREEZE	6850-01-464-9137	GL				
BRAKE FLUID,AUTOMOT	9150-01-102-9455	GL				
LUBRICATING OIL,ENG	9150-01-177-3988	QT				
GREASE,AUTOMOTIVE A	9150-01-197-7693	CA				
HYDRAULIC FLUID, FRH	9150-00-111-6256	QT				
LUBRICATING OIL,ENG 30WT	9150-01-460-7526	QT				
LUBRICATING OIL,ENG 15/40	9150-01-438-6076	QT				

CLASS IV

TYPE PACKAGE	Description	O/H Balance (Number of Packages)	Required Quantity	Time/Date Required	Total Short Tons
A	60 Rolls Concertina, 2 rolls barb wire, 160 6' pickets, 600 sand bags, 6 sets of sector stakes, 12 grazing logs, 6 sets of instruction, 77 landscape timbers, 6 sheets of plywood wood, 4 wooden pallets, 3 boxes of nails				
B	30 rolls of concertina, 160 6' pickets				
C	77 Landscape timbers				
D	30 sheets of plywood				
E	10 rolls of barbwire				

CLASS V

TYPE OF AMMUNITION	DODIC	UBL (Rounds)	OH Balance (Rounds)	OR%	Projected Requirement Next 24 hours	Projected Requirement Next 48 hours	Projected Requirement Next 72 hours
CTG. 9MM BALL M882	A363	100	0	● 0.1			

CLASS VII

TYPE OF EQUIPMENT	AUTHORIZED /REQUIRED	On-Hand	FMC/OR	OR%	12 Hour Projected OR%	24 Hour Projected OR%	QTY in DS MAINT or Below	Equipment Evacuated to GS
Ground Systems								
M1121/M966	10	10	7	70.0%				
M1114/M1151	100	100	72	72.0%				
M1025/26	100	100	100	100.0%				
M1097A2 CONTACT TRK	100	100	100	100.0%				
M1097A2 SHELTER S788	100	100	100	100.0%				
M707 KNIGHT	4	4	4	100.0%				
M1113 HVY HMMWV	100	100	100	100.0%				
M1038 HMMWV	100	100	100	100.0%				
M1090 / M1090A1WW MTV DUMP	100	100	100	100.0%				
FLU 419 SEE	100	100	100	100.0%				
M105 DEUCE	100	100	100	100.0%				
MW24C LOADER	100	100	100	100.0%				
M897 AMBULANCE	100	100	100	100.0%				
M1076 LHS TRAILER	100	100	100	100.0%				
M1120 LHS	100	100	100	100.0%				
M10A/10000M 10K FORKLIFT	100	100	100	100.0%				
M978 HEMMT FUELER	100	100	100	100.0%				
TWPS WATER PURIFICATION	3	3	2	66.7%				
M1078 LMTV CGO	100	100	100	100.0%				
M1083 HMTV CGO	100	100	100	100.0%				
M984A1 HEMMT WRECKER	100	100	100	100.0%				
M1089WW HMTV WRECKER	100	100	100	100.0%				

CLASS VII

TYPE OF EQUIPMENT	AUTHORIZED /REQUIRED	On-Hand	FMC/OR	OR%	12 Hour Projected OR%	24 Hour Projected OR%	QTY in DS MAINT or Below	Equipment Evacuated to GS
Small Arms & Crew Served Weapons								
M9 Pistol		25	25	100.0%				

CLASS VII

STAMIS	AUTH	On-Hand	FMC	OR%
VSAT	5	5	5	100.0%
BLUE FORCE TRACKER	100	100	75	75.0%
MTS	50	50	50	100.0%
SARSS-1	1	1	1	100.0%
SAMS-1	6	6	6	100.0%
SAMS-2	1	1	1	100.0%
PBUSE	20	20	20	100.0%
CAISI	35	35	35	100.0%

CLASS VIII

MEDICAL SUPPLIES Nomenclature/NSN	REQUIRED	AUTH	O/H	ESD

CLASS IX

SPARES Nomenclature/NSN	REQUIRED	AUTH	O/H	ESD

C-6

Commander's Assessment of Power and Capabilities		
Green	85-100%	Mission Capable
Amber	70-84%	Partially Mission Capable
Red	50-69%	Mission Ineffective
Black	below 49%	Requires Reconstitution

CLASS	Current Status	24 Hour Projection	48 Hour Projection	72 Hour Projection	Commander's Remarks
I	Green	Green	Yellow	Green	
II	Green	Green	Green	Green	
III (B)	Yellow	Green	Green	Green	
III (P)	Green	Green	Green	Green	
IV	Green	Green	Green	Green	
V	Green	Green	Green	Green	Insufficient resupply assets to meet RSR
VIII	Green	Green	Orange	Black	
VII	Yellow	Yellow	Yellow	Orange	Insufficient maintenance assets to repair damage. Losses
IX	Green	Yellow	Yellow	Orange	Demand for repair parts greater than PLL and ASL on-hand

Figure C-3: Example LOGSTAT

C-7

C - 4 Medical SITREP Format

Figure C-4 contains an example SITREP format.

Sample Medical SITREP Format
1. Current situation (significant changes in operational situation/planned or anticipated events next 24 hours)

2. DOD Health Service Support and Force Health Protection
   a) DOD population in Affected State
   b) DOD Population at Risk (PAR)
   c) DOD active duty medical units and grid location
   d) DOD reserve medical units and grid location

3. Medical operations (units and facilities)
   a) Bed availability and operational status (by type)
   b) Hospitalize patients by category
   c) All outpatient visits by category
   d) Class VIII (medical)
      i) Class VIIIA
      ii) Class VIIIB

4. Patient movement
   a) Required medical evacuation missions
   b) Medical evacuation missions conducted
   c) Dispositions
   d) Comments

5. Preventive medicine – occupational and environmental health
   a) Public health issues with mitigation recommendations
   b) Comments

6. Action Request Form/Mission Assignment Status   a) Comments

7. JTF Surgeon
   a) Priorities
   b) Current issues
   c) Future issues
   d) Additional critical information not addressed in SITREP
   e) Comments/remarks

8. POC for this report is (Name/Email/Phone#)

Figure C-4: Example SITREP

C - 5 Cost Worksheet

The following spreadsheet is used by the GCCs and the OSD Comptroller to capture costs associated with foreign disaster relief missions and may be used as a template at the JTF level.

DISASTER RELIEF OPERATION: DECEMBER 2010

Report Date:

* All figures are in thousands of dollars (000)

	Humanitarian Relief Supplies & Materials	1-Dec-10	2-Dec-10	3-Dec-10	4-Dec-10	TOTAL
	Medical Supplies					0.0
	Health & Comfort Packages					0.0
	Water & Water Storage					0.0
	Humanitarian Daily Rations					0.0
	All Other Humanitarian Relief Supplies					0.0
	Total Humanitarian Supplies & Materials	0.0	0.0	0.0	0.0	0.0

	Operational Support Costs	1-Dec-10	2-Dec-10	3-Dec-10	4-Dec-10	TOTAL
COSTS	Incremental Labor Costs (Includes Civilian Overtime and Contract Labor)					0.0
	Temporary Duty Costs					0.0
	Health Services, Clothing, & Misc Personnel Support					0.0
	Base Support (Billeting, mess, C4I, & other support for US forces)					0.0
	Airlift & Aviation Costs					0.0
	Sealift & Steaming Costs					0.0
	Port Handling & Misc Transportation Costs					0.0
	Other Operational Support Costs					0.0
	Total Operational Support	0.0	0.0	0.0	0.0	0.0

	Total Operation Costs (Humanitarian Supplies & Materials plus Operational Costs)	0.0	0.0	0.0	0.0	0.0

	Resource Authority	1-Dec-10	2-Dec-10	3-Dec-10	4-Dec-10	TOTAL
STATUS OF FUNDS	OHDACA Provided	0.0	0.0	0.0	0.0	0.0
	Less: Costs to Date	0.0	0.0	0.0	0.0	0.0
	Resources remaining after costs	0.0	0.0	0.0	0.0	0.0

Figure C-5: Cost Worksheet

This page intentionally left blank

UNCLASSIFIED

APPENDIX D - USEFUL FDR-RELATED WEBSITES

The following websites provide a variety of information related to Foreign Disaster Relief agencies and resources, as well as associated governmental and Non-Governmental Organizations (NGOs).

African Union
http://www.au.int/

The mission of the African Union Commission is to become an efficient and value-adding institution, driving the African integration and development process in close collaboration with African Union Member States, the Regional Economic Communities, and African citizens.

AlertNet
http://www.alertnet.org

Reuters Alertnet is a humanitarian news network that aims to keep relief professionals and the wider public up-to-date on humanitarian crises around the globe.

APAN
https://community.apan.org/

The All Partners Access Network (APAN) is an unclassified, non-dot-mil network providing interoperability and connectivity among partners over a common platform. The All Partners Access Network (APAN) fosters exchange and collaboration between the United States Department of Defense (DOD) and any external, country, organization, agency or individual that does not have ready access to traditional DOD systems and networks.

ASEAN (Association of Southeast Asian Nations)
http://www.aseansec.org/index2008.html

ASEAN's mission is to accelerate the economic growth, social progress and cultural development in the region through joint endeavors in the spirit of equality and partnership in order to strengthen the foundation for a prosperous and peaceful community of Southeast Asian Nations.

AusiAid
www.ausaid.gov.au

The Australian Agency for International Development (AusAID) is the Australian Government agency responsible for managing Australia's overseas aid program.

The objective of the Australian aid program is to assist developing countries reduce poverty and achieve sustainable development, in line with Australia's national interest.

AusAID's head office is in Canberra. AusAID also has representatives in 37 Australian diplomatic missions overseas.

CARE (Cooperative for Assistance and Relief Everywhere)
www.care.org

CARE is among the largest of the international NGOs and it is active in many disasters, notably Somalia. Additionally due to many long-term development programs, CARE workers are often on the ground and able to coordinate efforts for disaster relief.

CDC (Center for Disease Control)
http://www.cdc.gov

Center for Disease Control and Prevention within the Department of Health and Human Service; provides links to travel health, immunizations, and emergency preparedness.

CIA (Central Intelligence Agency) World Factbook
https://www.cia.gov/library/publications/the-worldfactbook

The content of the CIA World Factbook is in the public domain and includes information on the geography, people, government, economy, communication, transportation, military, and transnational issues of each country.

CIDI (Center for International Disease Information)
www.cidi.org

CIDI has information for individuals, groups, corporations, NGOs, embassies, media and others in order to provide more effective international emergency assistance.

ECHO (European Community Humanitarian Aid Office)
http://ec.europa.eu/echo/index_en.htm

ECHO was established in 1992 by the Second Delors Commission. Funding from the office affects 18 million people every year in 60 countries. It spends €700 million a year on humanitarian projects through over 200 partners (such as the Red Cross, Relief NGOs and UN agencies). It claims a key focus is to make EU aid more effective and humanitarian. (With the European Community being abolished in 2009, the office began to be known as the Humanitarian Aid department of the European Commission or European Union, but kept its ECHO abbreviation.)

European Union
http://europa.eu/index_en.htm

An economic and political union of 27 member states which are located primarily in Europe, the European Union ensures the free movement of people, goods, services, and capital in Europe; enacts legislation in justice and home affairs, and maintains common policies on trade, agriculture, fisheries and regional development.

FEWS NET (Famine Early Warning System Network)
http://www.fews.net

The goal of FEWS NET is to strengthen the abilities of African countries and regional organizations to manage the risk of food insecurity through the provision of timely and analytical early warning and vulnerability information.

HarmonieWeb
http://www.harmonieweb.org/Pages/Default.aspx

HARMONIEWeb is a site that allows government and non-government organizations to work in a collaborative environment to achieve common goals in the areas of Humanitarian Assistance, Disaster Response, and Stability and Reconstruction. Users can request portal sites to meet the collaborative needs of a given situation. Once the site is created, users build the sites, manage access, provide content, and designate administrators or site owners.

ICRC (International Committee of the Red Cross)
http://www.icrc.org/eng

The English website of the ICRC with information about the organization, international law (Geneva Conventions) and ICRC programs around the world.

ICVA (International Council of Voluntary Agencies)
http://www.icva.ch

Based in Geneva, ICVA is the primary hub of European NGOs.

IFRC (International Federation of the Red Cross and Red Crescent)
http://www.ifrc.org

Website of the IFRC Societies with access to information on IFRC programs, humanitarian principles and the NGO Code of Conduct.

InterAction
http://www.interaction.org

Based in Washington, DC, InterAction is the main hub of United States NGOs.

IRIN
www.irinnews.org

IRIN is a humanitarian news network that aims to keep relief professionals and the wider public up-to-date on humanitarian crises around the globe.

Logistics Cluster
http://www.logcluster.org

A field-based interagency logistics information platform to coordinate the widest possible participation among all humanitarian logistics actors (UN and NGO alike).

MCAST (Maritime Civil Affairs and Security Training Command)
http://www.mcast.navy.mil

The Maritime Civil Affairs and Security Training (MCAST) Command mans, trains, equips and deploys Sailors to facilitate and enable a Navy Component or Joint Task Force Commander to establish and enhance relations between military forces, governmental and nongovernmental organizations, and the civilian populace. Accomplished in a collaborative manner across the spectrum of operations in the maritime environment, MCAST Command executes civilian to military operations and military to military training, as directed, in support of security cooperation and security assistance requirements.

NATO
http://www.nato.int/cps/en/natolive/index.htm

The North Atlantic Treaty Organization is a military alliance whose mission is to promote peace, security, and democratic values through consultation and cooperation on defence and security issues.

OneResponse
http://oneresponse.info

OneResponse is an OCHA-managed, collaborative, inter-agency website designed to enhance humanitarian coordination within the cluster approach and support the predictable exchange of information in emergencies at the country level. The website will support the Clusters and OCHA fulfill their information management responsibilities as per existing IASC guidance.

ReliefWeb
http://www.reliefweb.int

OCHA's global hub for time-critical humanitarian information on complex emergencies and natural disasters.

Save the Children
http://www.savethechildren.org

Save the Children is the leading independent organization creating lasting change in the lives of children in need in the around the world. When disaster strikes around the world, Save the Children is there to save lives with food, medical care and education and remains to help communities rebuild through long-term recovery programs.

The Sphere Project
http://www.sphereproject.org/

The Sphere Project is an initiative to define and uphold the standards by which the global community responds to the plight of people affected by disasters, principally through a set of guidelines that are set out in the Humanitarian Charter and Minimum Standards in Disaster Response (commonly referred to as the Sphere Handbook). Sphere is based on two core beliefs: first, that those affected by disaster or conflict have a right to life with dignity and therefore a right to protection and assistance, and second, that all possible steps should be taken to alleviate human suffering arising out of disaster and conflict.

A digital copy of the Sphere Handbook can be found
http://www.sphereproject.org/component/option,com_docman/task,cat_view/gid,17/Itemid,203/lang,english/

State/CRS (Coordinator for Reconstruction and Stabilization)
http://www.state.gov/s/crs and http://www.crs.state.gov

Public access and password-protected websites for the Office of the Secretary of State's CRS with information on stabilization and reconstruction planning and operations, including the Essential Tasks Matrix.

State/PRM (Population, Refugees, and Migration)
http://www.state.gov/g/prm

Website for the Department of State's Bureau of Population, Refugees and Migration with information on USG programs for refugees and conflict victims.

USACAPOC (A) (United States Army Civil Affairs and Psychological Operations Command, Airborne)
http://www.usacapoc.army.mil/

USACAPOC (A)'s mission is to organize, train, equip and resource Army Reserve Civil Affairs and Psychological Operations forces for worldwide support to regional combatant commanders and other agencies as directed.

United Nations
http://www.un.org/en/

The UN strives to achieve international co-operation in solving international problems of an economic, social, cultural, or humanitarian character, and in promoting and encouraging respect for human rights and for fundamental freedoms for all without distinction as to race, sex, language, or religion.

UNDP (United Nations Development Program)
http://undp.org

UNDP is the UN's global development network, an organization advocating for change and connecting countries to knowledge, experience and resources to help people build a better life. UNDP leads the Early Recovery Cluster.

UN FAO (UN Food and Agriculture Organization)
http://www.fao.org

The UN's FAO leads international efforts to defeat hunger. Serving both developed and developing countries, the FAO acts as a neutral forum where all nations meet as equals to negotiate agreements and debate policy. FAO leads the Agriculture Cluster.

UNHCR (UN High Commissioner for Refugees)
http://www.unhcr.org/cgi-bin/texis/vtx/home

UNHCR safeguards the rights and well-being of refugees. The UNHCR leads the Camp Coordination Management and Emergency Shelter Clusters.

UNICEF (UN Children's Fund)
http://www.unicef.org

UNICEF is mandated to advocate for the protection of children's rights and help meet their basic needs. UNICEF works in 191 countries on improving health, nutrition, education, and reducing exploitation of children, including child soldiers. UNICEF co-leads the Education Cluster (along with Save the Children-UK), as well as leads the Water Sanitation and Hygiene (WASH) Cluster and Nutrition Cluster.

Unified Combatant Commands

US Africa Command (USAFRICOM)
http://www.africom.mil/

US Central Command (USCENTCOM)
http://www.centcom.mil/

US European Command (USEUCOM)
http://www.eucom.mil/

US Northern Command (USNORTHCOM)
http://www.northcom.mil/

US Pacific Command (USPACOM)
http://www.pacom.mil/

US Southern Command (USSOUTHCOM)
http://www.southcom.mil/appssc/index.php

UN OCHA (UN Office for the Coordination of Humanitarian Affairs)
http://ochaonline.un.org/

OCHA mobilizes and coordinates humanitarian assistance delivered by international and national partners. OCHA is headed by the Emergency Relief Coordinator (ERC), also titled Under-Secretary-General (USG), who is responsible for oversight of all emergencies requiring UN humanitarian assistance. OCHA co-leads the Emergency Telecommunications Cluster. OCHA is the process owner of the Emergency Telecommunications Cluster, providing overall coordination, preparedness and initial response.

UN OCHA Civil-Military Coordination Section
http://ochaonline.un.org/webpage.asp?Page=665

OCHA's Civil-Military and Coordination Section (CMCS) website allows access to the Oslo Guidelines on the Use of Military and Civil Defense Assets in Disaster Relief and other civil-military guidelines and information on CMCS UN Humanitarian Civil-Military Coordination (UN-CMCoord) training.

UN OCHA/UNDAC
http://ochaonline.un.org/OCHAHome/AboutUs/Coordination/UNDACSystem/tabid/5963/language/en-US/Default.aspx

OCHA's United Nations Disaster Assessment and Coordination (UNDAC) Team is a stand-by team of disaster management professionals. Upon request of a disaster-stricken country, the UNDAC team can be deployed within 24 hours to carry out rapid assessment of priority needs and support national authorities and the UN Resident Coordinator to coordinate international relief on-site. The website allows access to the UNDAC Handbook (similar to the United States Agency for International Development Field Operations Guide), information on UNDAC training, and a link to the International Search and Rescue Advisory Group (INSARAG).

United States Agency for International Development (USAID)
http://www.usaid.gov

Lead US agency for US response operations in foreign disaster response or humanitarian assistance.

The USAID Field Operations Guide (FOG) can be found at: http://www.usaid.gov/our_work/humanitarian_assistance/disaster_assistanc e/resources/pdf/fog_v4.pdf

United States Agency for International Development/Office of US Foreign Disaster Assistance (USAID/OFDA)
http://www.usaid.gov/our_work/humanitarian_assistance/disaster_assistance/

USAID/OFDA is the lead agency, and coordinator, for, USG responses to disasters in foreign countries. USAID/OFDA works to minimize the human costs of displacement, conflicts, and natural disasters. As the largest bilateral donor of humanitarian assistance, the USG has a unique role to play in shaping the nature of the assistance environment. Through close cooperation with other US Government agencies, bilateral and multilateral donors, host governments and implementing partners, USAID/OFDA's activities help maintain good relations and contribute positively to the United States' image abroad.

Virtual OSOCC (On-Site Operations Coordination Centers)
http://ocha.unog.ch/virtualosocc

Provides access to OCHA's Virtual On-Site Operations Coordination Centers (VOSOCC), which allow for real-time sharing of information related to specific emergencies and operations.

WFP (World Food Program)
http://www.wfp.org/

The United Nation's WFP is the world's largest humanitarian agency that fights hunger worldwide. The WFP leads the Logistics and Emergency Telecommunications Clusters.

D-8

WHO (World Health Organization)
http://www.who.int/en/

WHO is the directing and coordinating authority for health within the United Nations system. It is responsible for providing leadership on global health matters, shaping the health research agenda, setting norms and standards, articulating evidence-based policy options, providing technical support to countries and monitoring, and assessing health trends. WHO leads the Health Cluster.

This page intentionally left blank

APPENDIX E - TRAINING

Course Title: Joint Humanitarian Operations Course

Course Descriptor: As the lead agency for organizing US Government (USG) international disaster response, USAID's Office of US Foreign Disaster Assistance (USAID/OFDA) created the Joint Humanitarian Operations Course (JHOC) to establish a formal learning environment for select US military leaders and planners to discuss the relationship between USAID, its partners, and the US military, and to prepare participants to work collaboratively during Foreign Disaster Relief (FDR) operations.

To register: Email: mlu@usaid.gov, or contact USAID, Office of US Foreign Disaster Assistance, 529 14th Street NW, Suite 700, Washington, DC 20045, Office (202) 661-9382; Fax (202) 347-0315

Course Title: Humanitarian Assistance Response Training

Course Descriptor: The Humanitarian Assistance Response Training (HART) course provides military and response professionals with a two to four-day operational-level training course with practical information and tools for use in supporting civilian-led humanitarian assistance operations, including disaster response operations. The program is focused on civilian-military relations, including interacting with agencies of the Affected State and humanitarian agencies.

To register: http://coe-dmha.org/

Course Title: Interagency Logistics

Course Descriptor: This seminar style course offers a whole-of-government perspective on disaster relief logistics and is sponsored by DHS/FEMA. The Interagency Logistics Course (ILC) focuses on National and International level logistics operations by providing military and civilians stakeholders with insights into interagency logistics planning and execution.

To register: http://www.almc.army.mil/ALU_COURSES/8AF43551F35-MAIN.htm

Registration Assistance: Contact: ALU Registrar; Phone: (804) 765-4152/4149/4122

Course Title: Joint, Interagency, and Multinational Planner's Course (JIMPC)

Course Descriptor: The Joint Interagency Multinational Planner's Course (JIMPC) is a specialized short course addressing the dynamic challenges confronting mid-grade civilian and military planners who conduct interagency coordination for complex contingencies overseas. The five day

E-1

long course educates officers in the transforming organizations and processes that are being developed to improve a whole-of-government comprehensive approach to solving complex contingencies. This course educates officers in the latest developments in interagency coordination and serves as a forum for an exchange of best practices.

To register: http://www.jfsc.ndu.edu/schools_programs/jimpc/default.asp

Course Title: Joint Logistics (CLL-016)

Course Descriptor: This module provides professionals with knowledge of functional assignments that involve joint planning, inter-Service, and multinational logistics support, as well as joint logistics in a theater of operations. By completing this module, professionals will recognize the important roles and responsibilities within the joint logistics environment; the capabilities that joint logistics delivers; the important factors related to planning, executing, and controlling joint logistics; and the factors that will ensure a successful future for joint logistics.

To register: http://www.dau.mil/default.aspx (Continuous Learning Modules)

Training Provider: Joint Knowledge On-line

Programs: (See recommended courses listed below)

Website: http://jko.jfcom.mil to register for training:

1.   Go to http://jko.jfcom.mil and click "new user" *note if you have am AKO/DKO account skip to step 3.

2.   Register for an AKO/DKO account.

    a.   Military or Government civilian: Select "Joint Account" and follow the instructions for establishing an account. You must enter an external email address (.mil/.gov/company) to ensure delivery of your account information. Joint accounts become active after AKO/DKO verifies username usually within a few hours after submission. For questions regarding your Joint account contact the AKO Help Desk at 1-866-335-ARMY.

    b.   Contractors and other employee types: Select "Sponsored Account" and follow instructions for establishing an account.

        (1)   Once prompted for a sponsor information, enter "joint.training".

        (2)   You must enter an external email address (.mil/.gov/company) to ensure delivery of your account information. You will receive two emails notifications: the first will confirm your sponsor

approval (allow up to 24-hours), the second will confirm account adjudication by AKO/DKO (allow up to 24-hours after sponsor approval).

(3)   For questions and renewal information regarding your sponsored account, contact the JKO Help Desk at 1-757-203-5654.

3.   Once your account is established, go to http://jko.jfcom.mil and click "Enter JKO" you will be directed to the AKO/DKO login page. Login and you will be redirected to the JKO home page.

4.   For questions please contact the JKO helpdesk at 1-757-203-5654 or jkohelpdesk@jfcom.mil

Course Title: J3O P-US272-Department of Defense (DOD) 101 - Interagency Course-Section 001

Course Descriptor: The purpose of this course is to educate and inform individuals in the Department of Defense (DOD) and other US Government Agencies on the fundamental workings of DOD as an entity within the US Government and how DOD works within the interagency process. The student will become familiar with the key tasks and organizations within DOD and their interagency comparisons. At the conclusion of the course, the student will understand how DOD is organized and functions and its support to the interagency process.

Course Title: J3O P-US784-Department of Health and Human Service 101 - Interagency-Section 001

Course Descriptor: The purpose of this course is to educate and inform individuals in the Department of Defense (DOD) and other US Government Agencies on the fundamental workings of the Department of Health and Human Services (HHS) and how it operates within the interagency process. The student will become familiar with the history, organization, functions, and roles and responsibilities of HHS. At the conclusion of the course, the student will understand how the HHS is organized; how HHS responds to national public health and medical emergencies; how it compares to DOD and other agencies in the US Government and how it functions within the interagency process and in support of reconstruction and stabilization activities.

Course Title: J3O P-US422-Department of Homeland Security 101 - Interagency Course

Course Descriptor: The purpose of this course is to educate employees of the Department of Homeland Security and other domestic and international partners on the Departments international security role and presence. This

E-3

course is to be offered in conjunction with a number of other Interagency 101 courses designed to provide a baseline understanding of each organization. The student will become familiar with the structure, mission, responsibility, and organization of the Department of Homeland Security as they relate to international activities.

Course Title: J3O P-US830-Department of Justice 101 - Interagency Course-Section 001

Course Descriptor: The purpose of this course is to educate and inform individuals in the Department of Defense (DOD) and other US Government Agencies on the fundamental workings of the Department of Justice and how it supports reconstruction and stabilization operations. The student will become familiar with the organization, functions, and roles and responsibilities of DOJ. At the conclusion of the course, the student will understand how the DOJ is organized and how it functions within the interagency process in support of reconstruction and stabilization activities.

Course Title: J30 P-US298 Department of State 101 Interagency

Course Descriptor: The purpose of this course is to educate and inform individuals in the Department of Defense (DoD) and other US Government Agencies on the fundamental workings of the Department of State (DOS) and how it operates within the interagency process. The student will become familiar with the history, organization, functions, and roles and responsibilities of DOS. At the conclusion of the course, the student will understand how the DOS is organized; how it compares to DoD and other agencies in the US Government and how it functions within both the interagency process and in coordination with DoD. This course is one of the eleven modules that constitute the Interagency 101 course.

Course Title: J3O P-US345-USAID 101 - Interagency Course

Course Descriptor: The purpose of this course is to educate and inform individuals in the Department of Defense (DOD) and other US Government Agencies on the fundamental workings of the US Agency for International Development (USAID) and how it operates within the interagency process. The student will become familiar with the history, organization, functions, and roles and responsibilities of USAID. At the conclusion of the course, the student will understand how USAID is set up and how it compares to DOD and other agencies in the US Government as well as how it functions within both the interagency process and with DOD. This course is to be offered in conjunction with the DOD 101-Interagency and Department of State 101-Interagency courses.

Course Title: J3O P-US012-Joint Interagency Coordination Group (JIACG) Course-Section 001

Course Descriptor: This course provides an overview of JIACG and its structure, describes the JIACG environment and its players, and explains the JIACG process.

Course Title: Maritime Engagement and Crisis Response

Course Descriptor: Provide students with a basic knowledge of concepts, terms, organizations, and planning considerations for crisis response and limited contingency operations with an emphasis on Stability Operations. Present students with the US perspective relative to Limited Contingency Operations and Crisis Response in general and peace operations and humanitarian assistance operations in particular. Prepare students to assume duties on a staff that may be involved in conducting, planning, or supporting peace operations and/or humanitarian assistance and disaster relief operations.

To register: http://ewtglant.ahf.nmci.navy.mil/courses/cin/J-9E-0002.html

Phone: (757) 462-4504 / 4505, DSN: 253-4504 / 4505; FAX: (757) 462-7343, DSN: 253-7343; E-mail questions regarding quotas or cancellations to ewtglant_quotas@navy.mil

Course Title: Multinational Logistics

Course Descriptor: This course provides an overview of multinational operations. It acts as a force multiplier by familiarizing students with logistics strategy, doctrine, theory, programs and processes in a multinational environment.

To register: http://www.alu.army.mil/ALU_COURSES/ALMCNL-MAIN.htm

Registration Assistance: Contact: ALU Registrar; Phone: (804) 765-4152/4149/4122**Course Title: United Nations (UN) Civil-Military Coordination (UN CMCoord) course**

Course Descriptor: Since 2004, the Center of Excellence in Disaster Management & Humanitarian Assistance has partnered with the United Nations Office for the Coordination of Humanitarian Affairs (UN OCHA) to facilitate OCHA's Civil Military Coordination (UN-CMCoord) course. The course works to build a network of civilian and military personnel that are trained in international civil-military coordination. The UN-CMCoord Course is designed to address the need for coordination between international civilian humanitarian actors, especially UN humanitarian agencies, and international military forces in an international humanitarian emergency. This established UN training plays a critical role in building capacity to facilitate effective coordination in the field by bringing together practitioners from the spectrum of actors sharing operational space during a humanitarian crisis and training them on UN coordination mechanisms and internationally recognized guidelines for civil military coordination.

To register: http://coe-dmha.org/

Training Provider: United States Institute for Peace (USIP)

Training Program: The Academy for International Conflict Management and Peacebuilding is the education and training arm of the United States Institute of Peace. The Academy offers practitioner-oriented courses at the Institute's headquarters in Washington and elsewhere, conducts conflict management workshops and training in conflict zones abroad, and makes many of its courses and other resources available online to professionals, teachers, and students around the world.

Website: http://www.usip.org/training/online/

Training Provider: ReliefWeb

Programs Offers training by theme and country.

Website: http://reliefweb.int/trainings

Training Provider: United Nations Peace Operations Training Institute

Course Description: The Peace Operations Training Institute provides globally accessible and affordable distance-learning courses on peace support, humanitarian relief, and security operations. We are committed to bringing essential, practical knowledge to military and civilian personnel working toward peace worldwide

Website: http://www.peaceopstraining.org/

APPENDIX F - REFERENCES

The following documents were reviewed or considered during development this handbook or may be of interest to readers.

F - 1 Executive Orders/Public Laws/Statutes

Executive Order 12966- Foreign Disaster Assistance, 14 July 1995.

Foreign Assistance Act of 1961 (Public Law 87-195, as amended; 22 United States Code (USC) § § 2151 et seq.).

10 USC § 401 Humanitarian and Civic Assistance.

10 USC § 402 Transportation for Humanitarian Relief Supplies.

10 USC § 404 Foreign Disaster Assistance.

10 USC § 407 Humanitarian Mine Action.

10 USC § 2557 Excess Nonlethal Supplies for Humanitarian Relief.

10 USC § 2561 The Funded Transportation Program.

F - 2 Instructions/ Regulations/ Directives

Chairman of the Joint Chiefs of Staff Instruction (CJCSI) 2700.01D, "International Military Agreements for Rationalization, Standardization, and Interoperability Between the United States, Its Allies and Other Friendly Nations," 30 October 10.

CJCSI 3121.01B, "Standing Rules of Engagement/Standing Rules for the Use of Force for US Forces," 13 June 2005.

CJCSI 3214.01C, "Military Support of Foreign Consequence Management Operations," 11 January 2008.

CJCSI 3500.01B, "Joint Training Policy for the Armed Forces of the United States," 31 Dec 1999.

CJCSM 3113.01A, *Theater Engagement Planning*, 31 May 2000.

CJCSM 3122.03A, *Joint Operation Planning and Execution System Volume II Planning Formats and Guidance*, 31 Dec 1999.

CJCSM 3500.05A, *Joint Task Force Headquarters Master Training Guide*, 1 September 2003.

CJCSM 3500.04C, *Universal Joint Task List*, 1 July 2002.

Department of the Army, Graphic Training Aid (GTA) 41-01-003, *Civil Affairs Foreign Humanitarian Assistance Planning Guide*, August 2009.

Department of the Army, *Working with the Office of US Foreign Disaster Assistance*, October 2007.

Department of the Army, Field Manual 3-07, *Stability Operations*, October 2008.

Department of the Navy, Operational Navy Instruction (OPNAVINST) 5080.2 (Draft), *Cooperation with Non-Governmental Organizations in Permissive, Non-Hostile Environments*.

Department of Defense Directive (DODD) 2205.3, "Implementing Procedures for the HCA Program," 27 January 1995.

DODD 3000.05, "Stability Operations," 16 September 2009.

DODD 4160.21M, "Defense Material Disposition Manual," 18 August 1997.

DODD 5100.01, "Functions of the Department of Defense and its Major Components," 21 December 2010.

DODD 5105.22, "Defense Logistics Agency (DLA), 17 May 2006.

DODD 5105.64, "Defense Contract Management Agency (DCMA)," 27 September 2000.

DODD 5105.65, "Defense Security Cooperation Agency (DSCA)," 31 October 2000.

DODD 5100.46, "Foreign Disaster Relief," 4 December 1975.

DODD 5145.4, "Defense Legal Services Agency," 15 December 1989.

DODD 6200.04, "Force Health Protection," 9 October 2004.

DODD 6050.7, "Environmental Effects Abroad of Major DOD Actions," 5 March 2004.

Department of Defense Instruction (DODI) 2205.02, "Humanitarian and Civic Assistance Activities," 2 December 2008.

DODI 4715.5, "Management of Environmental Compliance at Overseas Installations," 22 April 1996.

DODI 4715.8, "Environmental Remediation for DOD Activities Overseas," 2 February 1998.

DODI 4715.05-G, "Overseas Environmental Baseline Guidance Document," 1 May 2007.

DODI 6000.16, "Military Health Support for Stability Operations," 17 May 2010.

DODI 8220.02, "Information and Communications Technology Capabilities for Support of Stabilization and Reconstruction, Disaster Relief, and Humanitarian and Civic Assistance Operations," 30 April 2009.

Joint Publication (JP) 1, *Doctrine for the Armed Forces of the United States*, 20 March 2009.

JP 1-02, *Department of Defense Dictionary of Military and Associated Terms*, 31 July 2010.

JP 1-04, *Legal Support to Military Operations*, 1 March 2007.

JP 1-06, *Financial Management in Support of Joint Operations*, 4 March 2008.

JP 2-0, *Joint Intelligence*, 22 June 2007.

JP 3-0, *Joint Operations*, 22 March 2010.

JP 3-07.3, *Joint Tactics, Techniques, and Procedures for Peace Operations*, 17 October 2007.

JP 3-08, *Interagency, Intergovernmental Organization, and Nongovernmental Organization Coordination during Joint Operations (Vol. 1 and 2)*, 17 March 2006.

JP 3-10, *Joint Security Operations in Theater*, 3 February 2010.

JP 3-16, *Multinational Operations*, 7 March 2007.

JP 3-29, *Foreign Humanitarian Assistance*, 17 March 2009.

JP 3-33, *Joint Task Force Headquarters*, 16 February 2007.

JP 3-34, *Joint Engineer Operations*, 12 February 2007.

JP 3-35, *Deployment and Redeployment Operations*, 7 May 2007.

JP 3-50, *Personnel Recovery*, 5 January 2007.

JP 3-57, *Joint Doctrine for Civil Military Operations*, 8 July 2008.

JP 3-61, *Doctrine for Public Affairs in Joint Operations*, 25 August 2010.

JP 4-0, *Joint Logistics*, 18 July 2008.

UNCLASSIFIED

JP 4-02, *Health Services Support*, 31 October 2006.

JP 4-08, *Joint Doctrine for Logistics of Multinational Operations*, 25 Sep 2002.

JP 4-09, *Joint Distribution Operations*, 5 February 2010.

JP 4-10, *Operational Contract Support*, 17 October 2008.

JP 5-0, *Joint Operations Planning*, 26 December 2006.

JP 6-0, *Joint Communications System*, 10 June 2010.

Joint Standard Operating Procedures, *Volume IX: Knowledge Management*, 02 February 2011.

Joint Staff Instruction 3820.01E, "Environmental Engineering Effects of DOD Actions," 30 September 2005.

National Security Presidential Directive 44, "Management of Interagency Efforts Concerning Reconstruction and Stabilization," 7 December 2005.

Navy Tactics, Techniques, and Procedures (NTTP) 3-57.3, *Navy Humanitarian and Civic Assistance Operations*, November 2009.

NTTP 4-02.6, *Hospital Ships*, June 2004.

Navy Warfare Development Command TACMEMO 3-07.6-06, *Foreign Humanitarian Assistance/Disaster Relief Operations Planning*, May 2006.

Presidential Decision Directive 56, "Managing Complex Contingency Operations," 20 May 1997.

US Southern Command Regulation 380-10, *Foreign Disclosure Program (For Official Use Only)*, 1 August 2010.

United States Joint Forces Command (USJFCOM), *Commander's Handbook for the Joint Interagency Coordination Group*, 1 March 2007.

USJFCOM, J-7 Pamphlet (Version 1), *US Government Draft Planning Framework for Reconstruction, Stabilization, and Conflict Transformation*, 1 December 2005.

F - 3 Other

A Common Perspective: The Joint Doctrine, Education, and Training Newsletter. Vol. 2, Number 4 (2011).

Air, Land, and Sea Application (ALSA) Center. *Multi-Service Tactics, Techniques, and Procedures for Conventional Forces and Special Operations Forces Integration and Interoperability.* March 2010.

ALSA Center. *Airfield Openings.* May 2007.

All, Pamela, LTC Daniel Miltenberger, and Thomas G Weiss. *Guide to IGOs, NGOs, and the Military in Peace and Relief Operations.* Washington, DC: United States Institute of Peace Press, 2000.

"Asia-Pacific Regional Guidelines for the Use of Foreign Military Assets in Natural Disaster Response Operations" Version 7.9 (Draft). United Nations Office for the Coordination of Humanitarian Affairs, Civil-Military Coordination Section. 15 November 2009.

Bazerman, M. and M. Watkins. *Predictable Surprises: The Disasters You Should Have Seen Coming.* Cambridge: Harvard University Press, 2004.

Captain Bill, Brian and Major Jeremy Marsh. *Operational Law Handbook.* Charlottesville: US Army, 2010.

Binnendijk, Hans and Patrick M. Cronin, ed. *Civilian Surge: Key to Complex Operations.* National Defense University: December, 2008.

The Brookings Institute. *Protecting Internally Displaced Persons: A Manual for Law and Policymakers.* October 2008.

Cross Sector Language Guide. World Cares Center. February, 2011.

Defense Threat Reduction Agency. *Foreign Consequence Management Legal Deskbook.* Washington, DC: DTRA, 2007.

Dziedzic, Michale J and COL Michael K. Seidl. "Provincial Reconstruction Teams." *United States Institute of Peace Special Report.* Special Report 147 (2005): 2-15.

Foster, Erin. "Humanitarian Principles and Guiding Documents." Civil Military Fusion Center: 16 November 2010.

Gates, Robert. "Memorandum for Commander, US Pacific Command." Subject: Disaster Assistance to the Philippines." 22 October, 2010.

Guidance to Security Assistance Offices and Joint Task Forces on Foreign Humanitarian Assistance and Disaster Relief (HA/DR). USSOUTHCOM. 22 May 2007.

"Guiding Principles for Public-Private Collaboration for Humanitarian Action." United Nations.

HAST Toolkit (Draft). USPACOM.

Humanitarian Assistance/Disaster Relief Lessons from a Hurricane (For Official Use Only). Quantico: USMC Center for Lessons Learned, 2006.

Institute for Defense Analysis. *Worldwide Humanitarian Assistance Logistics Handbook* (Vol. 1-3): 22 March 2004.

Interagency Teaming to Counter Irregular Threats. December 2009.

"Interim Supplement to Joint Task Force—Port Opening (JTF-PO), Seaport of Debarkation (SPOD) Concept of Operations (CONOPS)." USTRANSCOM. Version 1.05, 01 August 2008.

International Assistance System Concept of Operations. United States Department of State. 1 October 2010.

International Federation of Red Cross and Red Crescent Societies. "Code of Conduct for the IRC Movement and NGOs in Disaster Relief," 2002.

Joint Center for Operational Analysis. Vol. XII, Issue 2: Summer 2010.

Joint Civil Information Management Tactical Handbook. Version 2.0, 30 April 2010.

Joint Civil Information Management Users Manual. 29 April 2011.

"Joint Distribution Center Transition Template." DPO Concept Implementation and Assessment (TCJ5/4-J). 31 October 2008.

Joint Logistics Strategic Plan, 2010-2014. Department of Defense.

"Joint Logistics White Paper." Department of Defense. 4 June 2010.

Lieutenant General Keen, P.K. et al. "Foreign Disaster Response: Joint Task Force—Haiti Observations." *Military Review.* November-December (2010): 85-96.

Lawry Lynn MD, MSPH, MSc, ed. *Guide to Nongovernmental Organizations for the Military.* Washington, DC: United States Department of Defense, 2009.

Liaison: A journal of Civil-Military Humanitarian Relief Collaborations. Vol. 4, issue 1.

Liaison: A journal of Civil-Military Humanitarian Relief Collaborations. Vol. 6, issue 1.

"Military Surface Deployment & Distribution Command (SDDC) Support Plan to USTRANSCOM Joint Task Force—Port Opening (JTF-PO) Concept of Operations." Military Surface and Distribution Command. Version 2.0 15 May 2009.

Mull, Stephen D. "Memorandum for Michael L Bruhn, Executive Secretary, Department of Defense. Subject: Request for DOD Disaster Assistance to the Philippines." 8 October 2010.

Mull, Stephen D. "Memorandum for Michael L Bruhn, Executive Secretary, Department of Defense. Subject: Request for Humanitarian Assistance in Uzbekistan." 22 June 2010.

Multi-Agency Guide for Logistics Cooperation. Version 0.8.5

Multinational Information Preparation of the Operational Environment (MIPOE) for Humanitarian Assistance and Disaster Relief (HADR) "How To" Handbook (Draft). USPACOM.

Multinational Force Standing Operating Procedures (MNF SOP): HA/DR Mission Extract. USSOUTHCOM. Version 2.5. January 2010.

COL Nang, Roberto. "Medical Rules of Engagement." Internal Memorandum, 24 March 2011.

National Defense University. *Interagency Management of Complex Crisis Operations Handbook.* January 2003.

National Defense Strategy of 2008.

National Military Strategy of the United States of America, 2011.

National Security Strategy of 2010.

Office for the Coordination of Humanitarian Affairs (OCHA). *Oslo Guidelines: Guidelines on the Use of Foreign Military and Civil Defence Assets in Disaster Relief.* Revision 1.1, November 2007.

OCHA. *United Nations Disaster Assessment and Coordination Field Handbook.* 2000.

Office of the Coordinator for Stabilization and Reconstruction. "Reconstruction and Stabilization: Civilian Response," December 2005.

Peace & Stability Operations Journal Online. Vol 1, Issue 1: October 2010.

Perito, Robert M, ed. *Guide for Participants in Peace, Stability and Relief Operations*. Washington, DC: United States Institute of Peace Press, 2007.

Quadrennial Defense Review, 2010.

"Role of DOD in Foreign Disaster Relief" (PowerPoint Brief). Office of the Secretary of Defense, Partnership Strategy & Stability Operations. July, 2010.

Schnelle, D. "Introduction to Natural Disasters." Cited in G. Ciottone (Editor). *Disaster Medicine*. Pittsburgh: Mosby Elsevier, 2006.

Siegel, Adam B. "Civil-Military Marriage Counseling: Can this Union be Saved?" *Special Warfare*. December (2002): 28-34.

Smith, Daniel B. "Memorandum for Lieutenant Colonel Alfredo Najera, USA, Acting Executive Secretary, Department of Defense. Subject: C-130s for Chile Earthquake Relief Efforts." 5 March 2010.

The Sphere Project. *Humanitarian Charter and Minimum Standards in Disaster Response*. 2004.

Tactical Commander's Handbook for Theater Security Cooperation. 2009.

United Nations High Commissioner for Refugees. *Handbook for Emergencies*, July 2007.

United States Agency for International Development (USAID). *Field Operations Guide for Disaster Assessment and Response*. November 2005.

(USAID)/Bureau for Democracy, Conflict, and Humanitarian Assistance (DCHA), Office of US Foreign Disaster Relief (USAID/OFDA). *Guidance for Disaster Planning and Response*. 2010.

USAID/OFDA 2010 Hurricane Season Preparedness and Response Plan. 8 June 2010

USAID/OFDA. *Joint Humanitarian Operations Course: Civil-Military Roles in International Disaster Response*. Military Liaison Unit, 2010.

United States Government Interagency Complex Contingency Operations Organizational and Legal Handbook. Charlottesville: United States Army, 2004.

USIP/PKSOI. *Guiding Principles for Stabilization and re-Construction*.

"US Military Support to International Humanitarian Relief Operations: Fiscal and Legal Constraints" (Draft). MAJ Bradford B. Byrnes

"USTRANSCOM Support to Foreign Humanitarian Assistance." USTRANSCOM TCJ5/4-S. November 2010.

Vickers, Mike. "Action Memo for Deputy Secretary of Defense. Subject: Request for DOD Disaster Assistance Support in Philippines."

COL Vohr, Alex. "Haiti Disaster Relief: Logistics is the Operation." Unpublished paper.

Wagner, David. "Legal Aspects of Joint Operations" (PowerPoint Brief). United States Southern Command.

Wong, Marcia. "Conflict Transformation: The Nexus between State Weakness and the Global War on Terror." Remarks at the 17th Annual National Defense Industrial Association and Special Operations/Low Intensity Conflict Symposium. Crystal City, VA. 13 March 2006.

This page intentionally left blank

APPENDIX G - ACRONYMS

AAR	After Action Report
ACCE	Air Component Coordination Element
ACDA	Arms Control & Disarmament Agency
ACE	Aviation Combat Element
ACSA	Acquisition Cross-Service Agreements
ADB	Asian Development Bank
ADCON	Administrative Control
ADVON	Advance Echelon
AE	Aeromedical Evacuation
AELT	Aeromedical Evacuation Liaison Team
AFAFRICA	US Air Forces, Africa
AFB	Air Force Base
AFDRU	Austrian Forces Disaster Relief Unit
AFNORTH	US Air Force, North
AFSOUTH	US Air Forces Southern
AIDS	Acquired Immune Deficiency Syndrome
AKO	Army Knowledge Online
ALO	Aviation Liaison Officers
ALSA	Air, Land, and Sea Application Center
AMC	Air Mobility Command
AO	Area of Operations
AOR	Area of Responsibility
APAN	All Partners Access Network
APOD	Aerial Port of Debarkation
APOE	Aerial Port of Embarkation
ARG	Amphibious Ready Groups
ARNORTH	US Army, North
ASD	Assistant Secretary of Defense
ASD/APSA	Assistant Secretary of Defense for Asia/Pacific Security Affairs
ASD/HD&ASA	Assistant Secretary of Defense for Homeland Defense and America's Security Affairs
ASEAN	Association of Southeast Asian Nations
ASL	Above Sea Level
ASPR	US Department of Health and Human Services/Assistant Secretary for Preparedness and Response
AST	Airfield Survey Team
ATO	Air Tasking Order
AU	African Union
CA	Civil Affairs
CAB	Combat Aviation Brigade

CAO	Civil Affairs Operations
CAPE	Cost Assessment and Program Evaluation
CARE	Cooperative for Assistance and Relief Everywhere
CAT	Civil Affairs Teams
CBRNE	Chemical, Biological, Radiological, Nuclear, and High-Yield Explosives
CCATT	Critical Care Air Transport Teams
CCIR	Commander's Critical Information Requirements
CCT	Combat Controller Teams
CDC	Center for Disease Control
CDR	Commander (Navy rank – Commander/O-5)
CE	Command Element
CEB	Combat Engineer Battalion
CH	Chaplain (special staff)
CIA	Central Intelligence Agency
CIDI	Center for International Disease Information
CIM	Civil Information Management
CISM	Critical Incident Stress Management
CJCS	Chairman of the Joint Chiefs of Staff
CJCSI	Chairman of the Joint Chiefs of Staff Instruction
CJCSM	Chairman of the Joint Chiefs of Staff Manual
CMCS	Civil-Military Coordination Section
CMI	Classified Military Information
CMO	Civil Military Operations
CMOC	Civil Military Operations Center
COA	Courses of Action
COCOM	Combatant Commands
COE	Center for Excellence in Disaster Management and Humanitarian Assistance
COL	Colonel
COM	Chief of Mission
COMCAM	Combat Camera
COMPACFLT	Commander, Pacific Fleet
COMSEC	Communications Security
CONOPS	Concept of Operations
CONPLAN	Concept Plan
CONUS	Continental United States
COP	Common Operational Picture
COR	Contracting Officer Representative
COS	Chief of Staff

COSC	Combat and Operational Stress Control
COTS	Commercial Off-The-Shelf
CRE	Contingency Response Group
CRG	Contingency Response Element
CRM	Composite Risk Management
CRT	Contingency Response Team
CRTS	Casualty Receiving and Treatment Ships
CSEL	Command Senior Enlisted Leader
CSH	Combat Support Hospitals
CSL	Cooperative Security Location
CST	Communication Support Team
CUI	Controlled Unclassified Information
DART	Disaster Assistance Response Team
DASC	Direct Air Support Center
DASP	Disaster Assistance Support Program
DATT	Defense Attaché
DCHA	Bureau for Democracy, Conflict, and Humanitarian Assistance (USAID)
DCM	Deputy Chief of Mission (American Embassy)
DCMA	Defense Contract Management Agency
DCST	Defense Logistics Agency Contingency Support Team
DDA	Designated Disclosure Authority
DEET	N,N-Diethyl-meta-toluamide (insect repellant)
DHCA	Democracy, Conflict, and Humanitarian Assistance (USAIC/OFDA)
DHS	Department of Homeland Security
DISA	Defense Information Systems Agency
DKO	Defense Knowledge Online
DLA	Defense Logistics Agency
DMT	Disaster Management Team (United Nations)
DOD	Department of Defense
DODD	Department of Defense Directive
DODI	Department of Defense Instruction
DOJ	Department of Justice
DOS	Department of State
DPARS	Delivery Performance Achievement and Recognition System
DR	Disaster Relief
DSCA	Defense Security Cooperation Agency
DTRA	Defense Threat Reduction Agency
EAP	East Asian and Pacific Affairs

ECHO	European Community Humanitarian Aid Office
EDRT	Expeditionary Disposal Remediation Team
EEFI	Essential Elements of Friendly Information
EMCA	Europe, Middle East, and Central Asia (USAID/OFDA regional team)
EMEDS	Emergency Medical Support
EP	Excess Property
EPU	Expeditionary Port Units
ER	Emergency Relief
ERC	Emergency Relief Coordinator
ESP	Engineer Support Plan
EU	European Union
EUR	European and Eurasian Affairs
EXECSEC	Executive Secretary (DOS)
EXORD	Execute Order
FAA	Foreign Assistance Act of 1961
FAO	Foreign Area Officer
FARP	Forward Arming and Refueling Point
FCC	Functional Combatant Commands
FDR	Foreign Disaster Relief
FEMA	Federal Emergency Management Agency
FEST	Forward Engineer Support Team
FEWS	Famine Early Warning System
FFIR	Friendly Force Information Requirements
FHA	Foreign Humanitarian Assistance
FHP	Force Health Protection
FOB	Forward Operating Base
FOG	Field Operations Guide
FP	Force Protection
FPCON	Force Protection Condition
FPP	Food for Peace Program
FSN	Foreign Service National
FST	Fuel Support Team
FY	Fiscal Year
GCC	Geographic Combatant Commander
GCE	Ground Combat Element
GEOINT	Geospatial Intelligence
GIG	Global Information Grid
GPC	Geospatial Planning Cell
GPMRC	Global Patient Movement Requirements Center

GPS	Global Positioning System
GSJFHQ	Global Standing Joint Force Headquarters
HA	Humanitarian Assistance
HA/DR	Humanitarian Assistance/Disaster Relief
HAA/M	Humanitarian Assistance Advisor/Military
HAP	Humanitarian Assistance Program
HART	Humanitarian Assistance Response Training
HAST	Humanitarian Assistance Survey Team
HAZMAT	Hazardous Material
HC	Humanitarian Coordinator
HCP	Health Care Provider
HDM	Office of Humanitarian Assistance, Disaster Relief, and Mine Action (DSCA/PGM/HDM)
HDTC	Humanitarian Demining Training Center
HHS	Health and Human Services
HIU	Humanitarian Information Unit
HMA	Humanitarian Mine Action
HOC	Humanitarian Operations Center
HQ	Headquarters
HSS	Health Service Support
IA	Interagency
IASC	Inter-Agency Standing Committee
ICAO	International Civil Aviation Organization
ICRC	International Committee of the Red Cross
ICVA	International Council of Voluntary Agencies
IDP	Internally Displaced Person
IDRA	International Disaster Relief Assistance
IFR	Instrument Flight Rules
IFRC	International Federation of the Red Cross and Red Crescent
IG	Inspector General
IGO	Intergovernmental Organization
IHC	International Humanitarian Community
IM	Information Management
INSARAG	International Search and Rescue Advisory Group
IO	International Organizations
IOM	International Organization for Migration
IPC	Interagency Policy Committee
IRIN	Integrated Regional Information Network
ISA	International Security Affairs
IT	Information Technology

UNCLASSIFIED

JAT	Joint Assessment Team
JCMEB	Joint Civil-Military Engineering Board
JCMOTF	Joint Civil-Military Operations Task Force
JCSE	Joint Communications Support Element
JDDOC	Joint Deployment Distribution Operations Center
JDT	Joint Deployable Team
JEC	Joint Enabling Capabilities
JECC	Joint Enabling Capabilities Command
JFACC	Joint Forces Air Component Commander
JFHQ	Joint Force Headquarters
JHOC	Joint Humanitarian Operations Course
JIACG	Joint Interagency Coordination Group
JIATF	Joint Interagency Task Force
JIC	Joint Information Center
JIIC	Joint Interagency Information Center
JKO	Joint Knowledge Online
JLLIS	Joint Lessons Learned Information System
JLOC	Joint Logistics Operations Center
JLOTS	Joint Logistics Over the Shore
JMD	Joint Manning Document
JOA	Joint Operations Area
JOC	Joint Operations Center
JOPES	Joint Operation Planning and Execution System
JOPP	Joint Operational Planning Process
JP	Joint Publication
JPASE	Joint Public Affairs Support Element
JPERSTAT	Joint Personnel Status
JPG	Joint Planning Group
JPMT	Joint Patient Movement Team
JPRA	Joint Personnel Recovery Agency
JPRC	Joint Personnel Recovery Center
JRSOI	Joint Reception, Staging, and Onward Integration
JS	Joint Staff
JTF	Joint Task Force
KMO	Knowledge Management Officer
LCB	Logistics Coordination Board
LCE	Logistics Combat Element
LES	Locally Engaged Staff
LFA	Lead Federal Agency

UNCLASSIFIED

LHA	Landing Helicopter Assault
LHD	Landing Helicopter Dock
LNO	Liaison Officers
LOC	Lines of Communication
LOD	Line of Duty
LOE	Lines of Effort
LOG	Logistics
LOGSTAT	Logistics Status
LON	Level of Need
LOTS	Logistics-Over-the-Shore
LPD	Landing Platform Dock
LSD	Landing Ship Dock
MA	Mortuary Affairs
MAGTF	Marine Air Ground Task Force
MARAD	Maritime Administration
MARCENT	Marine Forces, Central Command
MARFORAF	Marine Force, Africa
MARFORNORTH	Marine Force, North
MARFORPAC	Marine Force, Pacific
MARFORSOUTH	Marine Force, Southern
MASF	Mobile Aeromedical Staging Facility
MCA	Maritime Civil Affairs
MCAST	Maritime Civil Affairs and Security Training
MCDA	Military and Civil Defense Assets
MDRO	Mission Disaster Relief Officer
MEDEVAC	Medical Evacuation
MEDSITREP	Medical Situation Report
METOC	Meteorological and Oceanographic Officer
MEU	Marine Expeditionary Unit
MFM	Mass Fatality Management
MIA	Missing in Action
MIL	Military
MILGROUP	Military Group
MIPOE	Medical Intelligence Preparation of the Operational Environment (or) Multinational Information Preparation of the Operational Environment
MISO	Military Information Support Operations
MIST	Military Information Support Team
MITAM	Mission Tasking Matrix
MLE	Military Law Enforcement

UNCLASSIFIED

MLO	Military Liaison Officer (USAID/OFDA)
MLU	Military Liaison Unit (USAID/OFDA)
MOB	Main Operating Base
MOE	Measure of Effectiveness
MOP	Measure of Performance
MOU	Memorandum of Understanding
MP	Military Police
MRE	Meals Ready to Eat
MSC	Military Sealift Command
MST	Medical Strike Team
MTF	Medical Treatment Facility
MTT	Mobile Training Teams
MWR	Morale, Welfare, and Recreation
NATO	North Atlantic Treaty Organization
NAVAF	US Naval Forces, Africa Command
NAVELSG	Navy Expeditionary Logistics Support Group
NAVFAC	Naval Facilities Engineering Command
NAVSO	US Naval Forces Southern Command
NCF	Naval Construction Force
NCHB	Navy Cargo Handling Battalions
NCMI	National Center for Medical Intelligence
NEA	Near Eastern Affairs
NECC	Naval Expeditionary Combat Command
NGA	National Geospatial Intelligence Agency
NGO	Non-Governmental Organization
NIH	National Institutes of Health
NIPRNET	Non-classified Internet Protocol Router Network
NMCC	National Military Command Center
NOAA	National Oceanic & Atmospheric Administration
NSC	National Security Council
NSDD	National Security Decision Directive
NSPD	National Security Presidential Directive
NTM	National Technical Means
OAS	Organization of American States
OCHA	United Nations Office for the Coordination of Humanitarian Affairs
OCONUS	Outside the continental United States
OCS	Operational Contract Support
OFAC	Office of Foreign Assets Control
OFDA	Office of US Foreign Disaster Assistance, United States Agency

	for International Development (USAID/OFDA)
OHDACA	Overseas Humanitarian, Disaster, and Civic Aid
OMA	Office of Military Affairs
OPCON	Operational Control
OPNAVINST	Operational Navy Instruction
OPORD	Operations Order
OPT	Operational Planning Team
ORM	Operational Risk Management
OSC	Operational Stress Control
OSCE	Organization for Security and Cooperation in Europe
OSD	Office of the Secretary of Defense
OSOCC	On-Site Operations Coordination Centers
OTI	Office of Transition Initiatives (USAID)
PACAF	US Air Forces, Pacific (PACAF)
PACOM	US Pacific Command
PAG	Public Affairs Guidance
PAHO	Pan American Health Organization (WHO)
PAM	Preventive and Aerospace Medicine
PAO	Public Affairs Officer
PAR	Population at Risk
PAS	Public Affairs Section
PDHRA	Post-Deployment Health Risk Assessment
PE	Peace Enforcement
PERSTAT	Personal Status Report
PGM	Program's Office of Humanitarian Assistance, Disaster Relief, and Mine Action (DSCA/PGM/HDM)
PIR	Priority Information Requirements
PK	Peacekeeping
PM	Political-Military
POC	Point of Contact
POL	Petroleum, Oil, and Lubricants
POLAD	Political Advisor
POTUS	President of the United States
POW	Prisoner of War
PPE	Personal Protective Equipment
PR	Personnel Recovery
PRM	Bureau for Population, Refugees, and Migration (US Department of State)
PRT	Provincial Reconstruction Team
RAMCC	Regional Air Movement Coordination Center
RC	Resident Coordinator

G-9

UNCLASSIFIED

RFF	Request for Forces
RMT	Response Management Team (USAID/OFDA)
ROE	Rules of Engagement
RPOE	Rapid Port Opening Elements
RRF	Ready Reserve Force
RSO	Regional Security Officer
RST	Religious Support Team
S/ES-O/CMS	Operations Center's Office of Crisis Management Support (DOS)
SAR	Search and Rescue
SARO	Southern Africa Regional Office (USAID/OFDA)
SC	Strategic Communication
SCA	South and Central Asian Affairs
SCHR	Steering Committee for Humanitarian Response
SDA	Senior Development Advisor (OMA)
SDDC	Surface Deployment and Distribution Command
SECDEF	Secretary of Defense
SFA MTT	Security Force Assistance Mobile Training Teams
SG	Command Surgeon
SHAPE	Supreme Headquarters Allied Powers Europe
SIPRNET	Secret Internet Protocol Router Network
SITREP	Situation Report
SJA	Staff Judge Advocate
SME	Subject Matter Expert
SMT	Senior Management Team (USAID/OFDA)
SOCAF	US Special Operations Command, Africa
SOCPAC	US Special Operations Command, Pacific
SOCSOUTH	US Special Operations Command South
SOF	Special Operations Forces
SOFA	Status of Forces Agreement
SOLIC	Office of the Assistant Secretary of Defense for Special Operations and Low Intensity Conflict (ASD/SOLIC)
SOP	Standard Operating Procedure
SOUTHCOM	US Southern Command
SP	Security Police
SPEARR	Small Portable Expeditionary Aeromedical Rapid Response
SPINS	Special Instructions
SPM	Single Port Manager
SPOD	Seaport Point of Debarkation
SROE	Standing Rules of Engagement
SWAN	South, West, and North Africa (USAID/OFDA regional team)

G-10

TACC	Tactical Air Control Center
TACMEMO	Tactical Memorandum
TACON	Tactical Control
TALCE	Tactical Air Coordination Center
TASKORD	Tasking Order
TAV	Total Asset Visibility
TCC	Transportation Component Command
TF	Task Force
TISC	Transnational Information Sharing Cooperation
TOC	Tactical Operations Center
TPFDD	Time-Phased Force Deployment Data
TRANSCOM	US Transportation Command
TSCP	Theater Security Cooperation Plan
TSOC	Theater Special Operations Command
UAS	Unmanned Aerial Systems
UCC	Unified Combatant Commands
UISC	Unclassified Information Sharing Concept
UK	United Kingdom
UN	United Nations
UNDAC	United Nations Disaster Assessment and Coordination Team
UNDP	United Nations Development Program
UNHAS	United Nations Humanitarian Air Service
UNHCR	United Nations High Commissioner for Refugees
UNICEF	United Nations Children's Fund
USA	United States Army
USACAPOC	United States Army Civil Affairs and Psychological Operations Command
USACE	United States Army Corps of Engineers
USAF	United States Air Force
USAFRICOM	United States Africa Command
USAID	United States Agency for International Development
USARAF	United States Army, Africa Command
USARPAC	United States Army, Pacific Command
USARSO	United States Army South
USC	United States Code
USCENTCOM	United States Central Command
USDA	United States Department of Agriculture
USDP	Under Secretary of Defense for Policy
USEUCOM	United States European Command
USFF	United States Fleet Forces Command

G-11

USG	United States Government
USGS	United States Geological Survey
USIA	United States Information Agency
USIP	United States Institute of Peace
USJFCOM	United States Joint Forces Command
USMC	United States Marine Corps
USN	United States Navy
USNORTHCOM	United States Northern Command
USNS	United States Naval Ships
USPACOM	United States Pacific Command
USSOCOM	United States Special Operations Command
USSOUTHCOM	United States Southern Command
USSTRATCOM	United States Strategic Command
USTRANSCOM	United States Transportation Command
VDAP	Volcano Disaster Assistance Program
VEI	Volcanic Explosivity Index
VFR	Visual Flight Rules
VOSOCC	Virtual On-Site Coordination Center
WASH	Water, Sanitation, Hygiene
WFP	World Food Program
WG	Working Group
WHA	Western Hemisphere Affairs
WHO	World Health Association

READ ME FIRST

Do's:

- Apply humanitarian principles and guidelines to greatest extent possible
- Coordinate with USAID/OFDA and Country Team
- Send LNOs to USAID\OFDA\DART, the US Embassy, and appropriate UN clusters
- Request LNOs early from DOS, USAID/OFDA, and OGAs
- Rely upon the SJA to become familiar with legal authorities and funding limitations
- Plan for contingencies: no plan survives first contact with the disaster
- Whenever possible, plan to contract with local resources
- Develop a transition plan upon acceptance of a mission
- Be transparent in all activities with the Affected State
- Develop a common operating picture for all participants

Don't:

- Cause further harm to the affected population or the environment
- Contract for services without assessing the affect upon the local population
- Overwhelm Affected State infrastructure
- Provide unsustainable services, technology, or infrastructure
- Exaggerate or aggrandize the role of the US
- Give the impression that DOD is a limitless resource
- Think DOD is the best organization to perform every mission or that DOD can solve all the problems
- Volunteer to perform missions not tasked or coordinated with USAID
- Release information not properly approved by the Foreign Disclosure Officer or Public Affairs Officer
- Plan on a lot of bandwidth or use graphic intensive slides